ISO 9000
Quality System
Assessment Handbook

ISO 9000
Quality System
Assessment Handbook

David Hoyle

Butterworth-Heinemann
Linacre House, Jordan Hill, Oxford OX2 8DP
A division of Reed Educational and Professional Publishing Ltd

⍺ A member of the Reed Elsevier plc group

OXFORD   BOSTON   JOHANNESBURG
MELBOURNE   NEW DELHI   SINGAPORE

First published 1996
Reprinted 1997

British Library Cataloguing in Publication Data
A catalogue record for this book is available from the British Library

Library of Congress Cataloguing in Publication Data
A catalogue record for this book is available from the Library of Congress

ISBN 0 7506 2563 5

Produced by Butford Technical Publishing, Bodenham, Hereford
Printed and bound in Great Britain by Hartnolls Limited, Bodmin, Cornwall

Contents

Preface

Many companies have developed quality systems, often in response to customer pressure but sometimes in order to improve the effectiveness of their operations. Whatever the reason, such systems need to be effective, otherwise the expected benefits will not be gained. Audits are required by ISO 9000 as a means of establishing the effectiveness of quality systems and are used by certification bodies to determine whether such systems meet the criteria for ISO 9000 registration. So an understanding of quality system auditing is essential both for those producing goods and services and for those assessing companies that do.

Auditing is not an exact science and its effectiveness is in the hands of the auditors. Some approach the task like inspectors, bent on finding nonconformity, others see their task as investigators and attempt to assist companies improve their performance. There will be some auditors who regard the task as a chore and reveal little of any value to the company's management.

I was once told a story of an auditor and an archbishop who, having passed into the next world, arrived at the pearly gates to be greeted by St Peter. St Peter said that he had only one name on his list that day. The archbishop was first to respond by insisting that the name on the list must be his as he had devoted his life to God's work. St Peter was quick to point out that unfortunately, the archbishop's name was not on the list and, in fact, it was the name of the auditor, who had put the fear of God into more people in one day than the archbishop had done in a lifetime.

If you are an auditor and you wish to emulate the auditor in the story, then I might suggest you read some other text. If on the other hand you would like to help companies succeed and be remembered for the added value you imparted when conducting audits, then read on.

Allan Sayle writes[1] that 'the professionally conducted audit presents a true and fair view that management can rely on'. He continues by saying that 'the audit is the only management tool that can almost guarantee to find the real cause of business problems and offer permanent solutions for the prevention of avoidable costs'. Would that this were true. Unfortunately, as so often in life, things do not quite meet our expectations. But is everyone's expectation of an audit the same? Sayle writes from the perspective of an auditor who could be either employed by the organization being audited or by one of their customers.

In the field of third party certification audits, the objective is different and it is this and the observed behaviour of the auditors that might cause some people to question Sayle's idea of an audit. There is no doubt from my own experiences of auditing that one sees so many ways of doing things that auditors would naturally possess a wealth of information of use to those being audited. Ways of interpreting requirements that are less costly, ways of overcoming difficulties in meeting requirements and ways of developing reliable management systems. If this experience is used wisely, then the auditor does indeed provide value for money. But if the auditor is merely an instrument of measurement then, whilst important, the value of the audit is diminished considerably and is treated as either a necessary evil or a cosmetic exercise. Many auditors do not attempt to find the cause of the problems they discover, neither do they offer solutions. In fact it becomes a legal issue if the auditor is not an employee of the company audited, since responsibility rests with those who offer the advice not those who take it. If a Medical Practitioner told you to take a certain medicine and you became seriously ill as a result, you could legitimately sue the doctor. So it is not surprising that auditors don't offer solutions to a company's problems. In fact the new draft of EN 45012 makes it quite clear that the consultancy and assessment functions of certification bodies have to be financially and organizationally independent. However, it still does not prevent a person owning both companies, providing the assessment function is not subject to undue influence.

Sayle again writes in *Quality World*[2] condemning the present standard of auditing and auditors who have climbed aboard the ISO 9000 band-wagon. Whilst preparations for this book were nearly complete at this time, Sayle's article provides a persuasive reason why another book on auditing is needed, particularly one that addresses third party audits.

The intention has been to pack the book with information that will assist auditors do their job more professionally. Planning check lists, pre-audit visit check lists, meeting

[1] *Meeting ISO 9000 in a TQM World*, Allan J Sayle 1991

[2] *Quality World*, April 1995

check lists, auditing check lists, do's and don'ts, questioning techniques, examples of nonconformity statements, flow charts for every stage in the process, forms, audit report contents and much more. Whilst the book primarily addresses external audits, many of the aspects covered equally apply to internal quality audits. Readers will notice that the approach I advocate is process-based not element-based and therefore the audit check lists are designed around business processes and not elements of the standard so they may be quite unique, but they work for me.

I offer a word of caution when studying examples of the nonconformity statements included. They can easily be taken out of context on their own and views will differ on whether they are valid or invalid. The examples are intended only as illustrations. In particular audits, those which I regard as valid may be invalid and vice-versa. When reading the detail, it is easy to forget that the auditor's task is to gain confidence. Auditing is an assurance activity, which is why ISO 9001 is referred to as a quality assurance standard and ISO 9004 a quality management standard. If taken out of context, some of the points I have made may imply there is a right and wrong way of doing things, a right and wrong interpretation of the standard etc. There is no right and wrong in auditing, only best practice and confidence. I have tried to give options leaving the reader to choose which is appropriate, particularly when conducting the documentation audit, the pre-audit visit, reporting the audit findings and conducting surveillance. The practices of each certification body will differ slightly and although each has to satisfy EN 45012, there is much latitude in these requirements concerning the modus operandi.

The interpretation of ISO 9000 is the same as I covered in my *ISO 9000 Quality Systems Handbook*, which was written primarily for those implementing the standard. Much of the advice given in the former book is also relevant to auditors and therefore readers are referred to the Handbook for a more detailed study of the ISO 9001 requirements. Whilst my previous book covers the standard more thoroughly than this book, in some ways this book goes further by breaking down the requirements into lists and addressing the inconsistencies in ISO 9000 in a form that will help auditors avoid the many pitfalls when reporting their findings.

The book is addressed to auditors so why not title it *Quality System Auditing*, as many other books in the field? As will be seen in Chapter 1, in the field of quality auditing is a mine field of terminology. The process to which companies submit themselves is an assessment of their quality system, performed by auditors. The auditors gather facts and then assess the effectiveness of the system, hence the term Quality System Assessment and hence the title of the book. I have charted a course through this mine field of terminology in Chapter 1, but even with internationally-agreed definitions, understanding is not easily achieved, as many of these definitions use similar words. It's difficult enough for the fluent English-speaking nations. Having had personal experience in the non-English speaking countries, I know it is even more difficult for them.

The first six chapters cover the auditing process: first in outline, then a chapter on audit management, followed by four chapters following through the four phases of an audit. There follows another four chapters, each intending to assist auditors in some aspect of their job. The appendices include some statistics on ISO 9000, a cross reference to ISO 10011, the standard on Quality System Auditing, and lastly a dictionary of auditing terms placed at the back so it is easy to find.

Finally, I do not profess to be a national authority on auditing. The views expressed are my own. They may not be shared by other auditors. Whatever your persuasion, I hope you may find something of use in this book. If as a result of reading this book, you find opportunities for improvement then I will have achieved my objective. If you would like to quarrel with some of the points I make or purchase a disk containing further data, contact information is given at the end of the book. A second edition based on the wisdom of a thousand practising auditors will indeed be a valuable text.

David Hoyle

Monmouth, May 1996

Acknowledgements

Over the last five years I have carried out numerous audits and conducted both Internal Auditor and Lead Auditor training courses in many countries of the world, all on behalf of Neville-Clarke International. In particular I would like to acknowledge Bob Birtwhistle who provided the opportunities and Steve Meakin who provided the basic material from which I developed my own understanding of auditing practices. Also thanks to Lead Tutors Andrew Turnbull, Keith Hamlyn and Ken Johnson who provided an excellent training environment in which we learnt the best way to conduct audits and teach auditing skills.

I would also like to acknowledge the many contributors to 'Auditors Corner', the section of the IQA's monthly journal, Quality World, which I initiated to help both auditors and auditees. I have used parts of some contributions in my analysis of nonconformity definitions in Chapter 5.

Thanks to Chris Hawkins for providing the story in the preface and for helping me realize that software auditing is no different for someone with a design and development background.

I am grateful to John Cranston of LRQA and Tony Harper of DNV QA who provided critiques of the draft and helped with clarifying some of the variations in auditing practices.

A special thanks to Allan Sayle, whose excellent book *Management Audits* has provided a continual source of inspiration and whilst covering similar ground remains a complementary text.

I also acknowledge the British Standards Institution, Mobil Europe Limited, the Institute of Quality Assurance, the International Register of Certificated Auditors and the Supplier Quality Requirements Task Force of Chrysler/Ford/General Motors.

However, the book would not have been possible without the hundreds of student auditors who have passed through my training courses and provided such a wealth of experiences.

Chapter 1

Principles of auditing

Purpose of audits

All audits are performed to establish facts rather than faults. They aim to establish, by unbiased means, factual information on some aspect of performance. As the performance of any organization will degrade unless some checks are carried out, audits are performed as a safeguard against a deterioration in standards. Hence all audits will detect variation from pre-defined standards. Some variation may be tolerable, other variation may be unacceptable and if left to continue may well signify loss of business, credibility and customer confidence. The performance of an organization with respect to the quality of its products and service is one aspect which can be audited. Other aspects include financial, health, safety, the environment, contract compliance and technical capability. For audits to be useful they have to be against standards which the organization is committed to meeting, otherwise they will not be taken seriously. In the quality field such a standard is the ISO 9000 series but others may be defence standards, industry standards, specific contracts as well as organizational policies, standards and procedures.

Audit objectives

All audits aim to provide assurance to those who commissioned them and establish:

- Whether certain agreed provisions, if implemented will yield the required results

- Whether only the agreed provisions are being implemented

- Whether the provisions have yielded results that are fit for their purpose and meet the needs of those who require them

and as a result whether a certificate of conformance can be issued or whether some improvement is necessary before awarding such a certificate.

In aiming to verify certain aspects of performance, the objective of these activities is:

- To find opportunities for improvement in the management system (normally first party audits – see below)

- To certify products or services as conforming with specified requirements (normally first and second party audits – see below)

- To approve an organization for the supply of products and services (normally second party audits – see below)

- To recommend registration or continued registration of the organization to a national or international management system standard (normally third party audits – see below)

Characteristics of audits

Audits should not change the performance of what is being measured, otherwise the facts gathered will not be representative of the true performance of the audit subject (the product, service, process or system). However, in quality system auditing, as a person is the medium through which information is acquired about a subject, it is virtually impossible to avoid changing the subject to some degree as it is unlikely that the auditee will act entirely as normal when questioned about the system. Quality system auditors therefore need to be particularly skilful in extracting information.

Audits should always be performed by someone who has no responsibility for what is being measured, otherwise they may bias the results. If the auditor does have some responsibility for the activities being examined then they could be blind to the deficiencies present or attempt to hide problems of which they are aware.

Audits should not be performed to find faults or to apportion blame although the results of the audit may well be used for such purposes.

Audits are performed to obtain a level of confidence rather than 100% certainty. An audit will reveal a level of confidence upon which decisions can be made, decisions that do not require 100% certainty, so activities, products, documents etc. can be sampled. Based on the results of samples taken during the audit, it can be established whether the results obtained are representative of the population from which the samples were taken.

Audits need to be carried out by personnel who have the capability to investigate aspects of performance pertinent to the decision to be taken and hence auditors need to have sufficient knowledge of the subject they are auditing. A superficial knowledge of the subject will cause auditors to overlook important details, ignore critical areas, risks and factors upon which successful performance depends.

Types of audit

Whilst the aim is common to all audits, their are several types of quality audit: quality system audits, management audits, product audits, procedure audits and process audits. In each case, the auditor uses a different method and involves different people in the organization. Quality system audits include an element of management audits but also extend to technical areas of the business where product is designed, produced, installed and serviced.

Management audits[1]

These audits include such elements as:

- The Strategic Quality Audit. To verify that the strategic plans of the organization address current and future legal, environmental, safety and market quality requirements (sometimes referred to as the President's Audits[2]).

- The Policy Audit. To verify that the documented policies promulgate the requirements of the market and the objectives of the business.

- The Organization Audit. To verify that the organization is structured and resourced to implement the policies and to achieve the stated objectives efficiently and effectively.

---

[1] Sayle classes all audits as management audits in his book *Management Audits*.

[2] Juran, *Quality Control Handbook 1988*, Section 8 and Sayle, *Management Audits*, Chapter 27

Product/service audits

These audits include such elements as:

- The planning audit. To verify that the organization's plans or proposals for supplying a product or service will, if properly implemented, result in product or service that complies with specified requirements.

- The conformance audit. To verify that the product being produced or service being delivered conforms to specified requirements.

Process audits

These audits include such elements as:

- The planning audit. To verify that the plans for operating a result producing process will, if properly implemented, yield product which consistently meets the agreed specifications.

- Capability audit. To verify that a process has the capability to consistently yield product which meets agreed requirements.

Procedure audits

Procedure audits include such elements as:

- Documentation audit. To verify that the documented practices implement the approved policies and the relevant requirements of the standard and will, if properly implemented, provide an adequate degree of control over the organization's operations.

- Implementation audit. To verify that the activities and related results implement the approved documented practices.

System audits

These audits includes such elements as:

- Documentation audit. To verify that the documented system complies with the relevant requirements of the governing standards. They are also used to verify that the documented system provides the organization with the capability of supplying particular products and services that achieve certain specified requirements.

- Implementation audit. To verify that the activities and related results comply with the documented requirements and that the system is effective in providing an adequate degree of control over the organization's operations.

- System surveillance. To verify that the organization has maintained its quality system and it continues to be suitable for achieving its stated objectives and is effective in providing an adequate degree of control over the organization's operations.

Categories of audits

There are several categories of audits which can be applied whatever the audit objective.

First party audits

These are audits of a company or parts thereof by personnel employed by the company. These audits are also called *Internal Audits*.

Second party audits

These are audits carried out by customers upon their suppliers[3]. These audits are *External Audits* and are also called *Vendor Audits*. When second parties audit a supplier's quality system such audits are referred to as *Supplier Approval Audits* or *Quality System Assessments*[4].

Third party audits

These are audits carried out by personnel who are neither employees of the customer nor the supplier and are usually employees of certification bodies. They may be consultants or other persons or organizations having no direct interest in the supplier or the customer. Most third party audits are carried out by registrars or certification bodies which are themselves accredited to issue certificates to organizations that comply with specified requirements. These audits are also *External Audits* and can be referred to as *Certification Audits, Compliance Audits* and *Quality System Assessments*.

---

[3] A supplier is an organization that supplies products and/or services to a customer.

[4] The term Quality System Assessment is used in connection with assessments to QS 9000.

There is a view that audits by certification bodies that are under contract from the first party are not in fact true third party audits. They have no responsibility for the activities they audit so are in fact independent but are not wholly impartial since they are being paid by the organization they are auditing. It is claimed that they have a vested interest in the results. Were they to be too tough they may lose the contract. It is also claimed that certification bodies are lenient on the initial audit and then become tougher on the surveillance audits. In order to be competitive, certification bodies may undercut their rivals with cheap initial audits. The only safeguard at present is the independent inspection of the certification bodies by the Accreditation agency.

Category	Funding	Planning	Conduct	Reporting	Consequences
First Party	Supplier	Audit schedule covering quality system. Advanced notice of audit. Informal contact.	Informal – primarily because auditor knows auditee. Guides normally unnecessary.	Generally limited to opportunities for improvement. Advice given on corrective actions.	Either staff have to conform to the requirements or the requirements have to be changed.
Second Party	Customer	Audit plan and formal contact – may be unannounced or scheduled by deliverable or project phase.	Formal – tense and constrained by contract conditions. Guides necessary.	Formal. Good and bad points plus advice may be given on corrective action.	Loss of contract, removal from preferred suppliers list.
Third Party	Supplier	Audit brief, plan and formal contact – normally announced at regular intervals and scheduled on periodic basis.	Formal – Guides necessary.	Formal. Positive and negative points. Recommendation for certification but no specific advice is given on corrective action.	Can apply again or choose alternative third party. Possible loss of business if results made public.

Table 1-1 Comparison of audit categories

Terminology

In the auditing field many terms are used with similar meanings all of which can be confusing to the newcomer. Even experienced professionals will differ in their views as to the right term to use. Although a distinct advantage of the English language is its ability to provide a means of conveying the same message in many different ways, (a wonderful language for poets) this creates a distinct disadvantage when users intend to convey different messages by the words they use. In one context several words can have the same meaning when in another context each can have different meanings. If one person carries out an audit and another an assessment, are they doing the same thing or different things? The same applies for evaluations and appraisals. Terms assist in communicating concepts but if we apply different labels to the same concept we will fail to communicate properly. One man's audit is another man's assessment etc. It is a minefield through which it is difficult to chart a reliable course. That which follows is not intended to be definitive but intended to bring to the reader's attention the various terms, their similarities and application, so leaving them to decide the terms to use. It matters not what term people use providing that reliable communication results.

Audit, assessment or evaluation

External audits are often called assessments and internal audits, audits. Since its creation, the UK based Registration Board for Assessors (RBA) used the term 'assessment' but when the name changed to the International Register of Certificated Auditors (IRCA[5]), assessment became audit and assessors became auditors. The reason for the change was to align with ISO 10011 in which the term audit rather than assessment is used.

An audit is an examination of something to verify its accuracy. In old English it means a hearing of an account of some event. An assessment is an examination of something to determine what it is. Both words are nouns. In an assessment, information is passed to an assessor who applies certain rules and makes a judgement. In an audit, information is examined by the auditor against certain standards for accuracy. A tax assessor will determine the amount you will pay on the basis of the information provided and a tax auditor will verify whether the information provided is accurate. Alternatively a vehicle could be assessed following an accident to determine the amount of damage and the insurance claim could be audited to verify that the claim only covers the damage incurred in the accident and nothing else. Therefore in a financial context the two terms have different meanings.

Another problem with the terms audit and assessment is that they are not interchangeable. The term continuous assessment is not the same as continuous audit. A student would not consider he/she was being audited during a course of study. An assessment centre for young offenders is not the same as an audit centre. The young offenders are awaiting a decision about their future. A list of assessed suppliers is not the same as a list of audited suppliers. The suppliers that have been assessed and included in the list have all met the assessment criteria. The list of audited suppliers will contain details

---

[5] The IRCA is the International Register of Certificated Auditors, part of the UK's Institute of Quality Assurance and based in London.

of suppliers regardless of whether they have met certain criteria. One can make an assessment of something but not make an audit of something.

The ISO 8402 definition of quality audit is very precise although one could attach other labels to the definition and not corrupt the meaning. ISO 8402 defines a quality audit as:

> *'A systematic and independent examination to determine whether quality activities and related results comply with planned arrangements and whether these arrangements are implemented effectively and are suitable to achieve objectives.'*

By dissecting this definition an analysis would reveal that quality audits:

- Have to be systematic i.e. not random but an orderly and thorough examination.

- Have to be independent i.e. not performed by someone responsible for the activities, processes or products being examined.

- Have to determine whether quality activities comply with planned arrangements i.e. those activities in an organization which affect the ability of an entity to satisfy stated or implied needs. They will include all activities in the quality loop[6].

- Have to cover related results i.e. not merely an examination of the activities but an examination of the results of performing these activities, which may include documents, records, products, processes, decisions etc.

- Have to determine compliance with planned arrangements i.e. the policies and generic procedures and the product of such policies and generic procedures, such as contracts, plans, specifications, drawings, specific procedures and instructions.

- Have to determine whether the arrangements are implemented effectively i.e. not only that arrangements are implemented but whether the result of implementing the arrangements is effective. Effectiveness is judged by whether the right things are being done, targets met, deliveries achieved, schedules met, throughput maintained etc.

---

[6] The quality loop is illustrated in ISO 9004-1 and includes all operations in the product/service life cycle from marketing through to product/service disposal. The quality loop is also referred to by Juran as the quality spiral in his *Quality Control Handbook*.

- Have to determine whether the planned arrangements are suitable to achieve objectives i.e. whether the provisions made by the supplier serve the organization's objectives. These objectives will be the quality objectives.

There is no definition of a quality assessment in either ISO 8402 or BS 4778. However, there is no doubt that a quality audit is much more than an exercise in verifying the accuracy of records. There is a great deal of judgement being used in the conduct of a quality audit. In the financial context, audits verify accuracy and if accounts are not accurate, then the client may be suspected of fraud. If the figures do not add up then the auditor has to determine where the money has been deposited. A product quality audit would in fact warrant an examination of records to determine their accuracy and may involve re-examination of the product to determine whether the results are repeatable. A quality system audit may involve an examination of product but not to certify the product. The product examination is carried out to test the effectiveness of the system and any product could be chosen as the sample. A quality system audit would also involve an examination of operations and records of such operations to determine whether the records were a true reflection of the operations and, in this sense, an audit is being conducted. However, three areas of judgement are necessary, one concerning the extent of compliance with the requirements of the standard, one concerning the effectiveness of the quality system to meet defined objectives and one concerning the recommendation for certification. A definition of the term 'assessment' does appear in QS 9000, the standard being used by Ford, Chrysler and General Motors in place of ISO 9001. In this publication, an assessment is an evaluation process including a document review, an on-site audit, an analysis and a report.

In ISO 8402, a quality evaluation is 'a systematic examination of the extent to which an entity (e.g. a quality system) is capable of fulfilling specified requirements'. It states that it is the same as a quality assessment and a quality appraisal.

It is interesting to note in this context that in 1972, BS 4891[7] (now withdrawn) mentions Review and Evaluation. AQAP-2[8]-1984 was entitled 'Guide for the *evaluation* of a contractor's quality control system for compliance with AQAP-1' rather than 'Guide to the *assessment* of a contractor's quality control system', and in AQAP-15[9]-1986, the term System Evaluation is defined as a systematic examination of the effectiveness of the contractor's system concerned.

---

[7] BS 4891 November 1972 *A Guide to Quality Assurance*

[8] AQAP-2 is a NATO Allied Quality Assurance Publication.

[9] AQAP-15 is a glossary of terms used in NATO quality assurance publications.

The activity we call quality system auditing can be three activities: audit, evaluation and assessment. The audit establishes the facts, the evaluation establishes the effects and the assessment decides the verdict.

Third party quality system audits should therefore be called quality system assessments by virtue of the outcome[10]. Quality system audits not carried out by certification bodies can still be regarded as assessments if judgement is made regarding the organization's potential for being registered.

First party audits are usually confined to auditing with the management review performing the evaluation role.

Second party audit may or may not encompass audit, evaluation and assessment and therefore may be called audits or assessments.

In this handbook, the term 'assessment' is used when referring to the complete process (see Figure 1-1) and 'audit' is used when referring to the fact finding activity.

Auditor, assessor, auditee, assessee

The person who carries out the audit is called an auditor. The person who carries out an assessment is called an assessor. The organization that is subject to audit is defined as the auditee[11] but the organization subject to an assessment is referred to as the client, supplier, contractor etc. not as assessee. There is no such word in the English language. The term auditee is also used to describe the person the auditor interviews rather than the organization being audited. This seems more logical; otherwise we have to call the person being interviewed, the interviewee. In a second party situation, the organization being audited is the supplier or subcontractor. In a third party situation the organization being audited may be the client or the client's supplier as it depend on which body commissioned the audit. Since the organization performing the audit is the certification body or registrar, then it would be sensible to refer to the person being interviewed as the auditee and the organization being audited as the company.

Anyone can carry out an audit – they do not have to be qualified. People who carry out audits in ISO 9000 registered organizations and certification bodies do need to be qualified. Auditors who carry out internal and second party audits who do not assess the results and make recommendations to others are auditors. Auditors who do make

---

[10]This approach is endorsed by QS 9000 and the *Quality System Assessment Guide* (QSA) in which an assessment comprises a documentation review and an on-site audit.

[11]As defined in ISO 10011 Part 1.

such recommendations are assessors. On an external third party audit, the Lead Auditor is an assessor as he/she makes a judgement on the basis of the findings. The members of the audit team are simply auditors. Therefore the correct term for the leader should be the Assessor but usage of the term has degraded its meaning and so the term Lead Auditor is now commonly used.

Quality system certification, capability approval or qualification approval

BS 4778 1991 defines certification as 'the authoritative act of documenting compliance with requirements'. BS 4778 1991 defines 'Capability Approval' as 'the status given to a supplier for a range of items or services for which it has been shown that their declared design rules, manufacturing processes and quality system are capable of producing items or services, the quality of which meets the specified requirements'. This definition raises a number of issues:

- Is a quality system audit meant to establish supplier capability?

- Is an ISO 9000 certificate of registration awarded for a particular scope of business and products/services an approval of capability?

- Is supplier capability established through an audit, an assessment, an appraisal or an evaluation?

Qualification approval is defined in BS 4778 as 'the status given to a supplier whose product has been shown to meet all the requirements'. If we replace the word 'product' with 'entity' then since an entity according to ISO 8402 can be a quality system, Qualification approval becomes identical to Certification. Therefore in the field of quality system auditing, certification, capability approval and qualification approval all have the same meaning.

Accreditation or certification

Accreditation signifies that a person or organization is authorized to certify other persons or organizations. Certification is performed by accredited and non-accredited organizations. When an organization is accredited it means that their organization has been examined against defined criteria and found capable of certifying certain organizations.

Certification bodies such as LRQA, BSIQA, SGS Yarsley, DNV QA, BVQI etc. are accredited only for a certain range of certification activities. This is largely limited by the qualifications and experience of the personnel they employ. They may request an extension to their accredited scope when they are able to demonstrate a capability in an area outside their existing scope of accreditation. For example, a certification body may be accredited for certifying steel producers but not cement producers. By acquiring the necessary skills and expertise and carrying out a number of non-accredited assessments of cement factories, they could apply to extend their scope of accreditation so that they could perform accredited assessments of cement producers. The difference between accredited and non-accredited relates to the competence of the auditors not the organization being audited. When a purchaser chooses a supplier, knowledge that the quality system certification was carried out by an accredited certification body should give more confidence in the quality of the product than if carried out by non-accredited bodies. This is not necessarily so but in principle, audits by accredited bodies are more reliable than those carried out by non-accredited bodies. In the UK this principle has been taken a step further in the software, pharmaceuticals and aerospace sectors where special rules are applied to distinguish between competent and non-competent auditors.

Companies may request an extension to their certificated scope when they are able to demonstrate a capability in an area outside their existing scope of certification. For example, a company may employ a certificated quality system to produce nylon fabrics. It later wishes to produce cotton fabrics. Since the processes are different the quality system has to be modified and requires recertification and an extension to the scope of certification.

Registrar or certification body

In the USA the organization that carries out quality system certification is called a *registrar*. In the UK and elsewhere, this organization is called a *certification body*[12]. The process of certifying a quality system is sometimes called *registration*. This is because once the auditors have completed their audit they may recommend that the organization is listed in a *register* of assessed organizations that meet the requirements of the ISO 9000 series.

Implementation audit, conformance audit or compliance audit

Conformity is defined in ISO 8402 as the fulfilment of specified requirements. A conformity audit could therefore be deemed to be an audit to determine whether specified requirements have been fulfilled. The term 'conform' means to have the same form. Therefore conformity would seem to relate to objects that have a form rather

---

[12] A certification body is a body that conducts certification of conformity. The European Standard EN 45012 contains general criteria for certification bodies operating quality system certification.

than systems, but it could relate to actions which have to conform to certain rules. Compliance is not defined in ISO 8402. 'Comply' means to act in accordance with a command or rule. There is little distinction between the two terms but in general products conform, people comply. For product quality audits the term conformance audit would be appropriate when verifying that the product fulfilled the specified requirements. For quality system audits, the term compliance audit would be appropriate when verifying that the actions of the organization met the requirements of the standard. This distinction is supported by AQAP-15 where it states that systems comply with requirements and products conform to specifications. The compliance audit is therefore another term for a quality system audit rather than an element of a quality system audit. An implementation audit is that part of a compliance audit that deals with the implementation of the documented system. Using the term implementation audit overcomes any confusion between products and systems and whether conformance or compliance audits cover both documentation and its implementation.

Documentation audit, adequacy audit or desk top audit

The concept to which these terms relate is the examination of the quality system documentation to determine the extent to which it complies with the requirements of the governing standard (ISO 9001, ISO 9002, ISO 9003 or a customer standard).

This activity is not the same as the preliminary review referred to in ISO 10011. The preliminary review is conducted as a basis for planning the audit in order to reveal whether the system is sufficiently developed to go ahead with the audit. Some auditors refer to this preliminary review as an adequacy audit since unless the documented system is adequate there is little point in proceeding with the audit. However, the term 'adequacy audit' is also given to the above concept which is confusing to other auditors.

An in-depth examination of the quality system documentation is necessary in addition to a cursory examination for audit planning purposes so the terms we use ought to make the distinction clear. Figure 1-1 depicts four elements of an audit of which one is the documentation element. It would therefore be logical to name the activity of examining the documentation as a Documentation Audit. The fact that the documentation audit does not involve interviews means that it can be performed sat at a desk, hence the name 'Desk Top Audit'. Other terms that have been used for the same activity are Systems Audit, Quality Manual Audit, Planning Audit. The one thing they all have in common is the concept of checking the documented provisions the company has made for meeting certain requirements before checking that the provisions have been implemented.

A Planner's Tale

Preparation of the audit was going well. Neil had acquired all the information he needed from the company – a map showing directions, the Quality Manual, a company . brochure and an organization chart. It impressed him even more when he saw that the Quality Manual responded element by element to ISO 9001.

Neil sat down and began to plan the audit. He had worked out how many auditors he needed as the company had indicated there were 190 staff at that particular location.

The company designed and manufactured computerized test equipment so when Neil examined the organization chart and the Quality Manual he assigned one auditor to cover design, one to cover purchasing and two to cover manufacturing. He was going to interview the CEO and the Quality Manager himself.

Neil presented his plan at the auditors' briefing. The other auditors appeared happy with the plan except for David. David wondered why Neil had allocated so much time to the manufacturing areas. Neil replied that it was a manufacturing company that designed its own products. David asked Neil if there was any information showing how many were employed in each department. Neil didn't know the answer to this question so undertook to find out.

Fifteen minutes later, Neil returned to the meeting room showing a little embarrassment. It transpired that of the 190 staff, 30 were employed in Manufacturing, 70 in Design, 5 in Quality Assurance, 15 in Purchasing and the remainder in Administration, Sales and Marketing. The Quality Manager had told Neil that their manufacturing department was entirely engaged in assembly and most of the parts were purchased off-the-shelf from component suppliers. Even the manufacture of the equipment cases was subcontracted. In the design department they employed 20 software programmers.

Admitting defeat, Neil had to revise his audit plan considerably and engage a software specialist.

The moral of this story is:

Nothing is quite what it seems at first sight!

Quality system assessments

So far we have covered the types and categories of audits and the terminology connected with quality auditing. However, the subject of this book is Quality System Assessment and, as will have been detected so far, quality system assessments are but one type of audit and can be performed by first, second or third parties. Also the labels are not sacrosanct so a quality system audit can be referred to as a Supplier Approval Audit or a Quality System Audit. This may be confusing but that is the nature of the terminology jungle. What we call things is relatively unimportant providing we know what activity we are discussing. If we are intent on establishing that a quality system is effective, then whatever we call it the process is the same, although the standards will vary (see later).

A quality system is a management tool to achieve, sustain and improve the quality of results. The results may be products, services and decisions which come out of result producing processes. The system consists of the organization, responsibilities, documentation, processes and resources to achieve, sustain and improve quality. Within companies, quality system audits are the measurement component of the quality system. Having established a quality system it is necessary to install measures that will inform management whether the system is being effective. Installing any system without some means of verifying whether it is doing the job it is intended to do is a waste of time and effort.

Externally the quality system assessment is performed to determine the capability of an organization to supply specified products and services[13]. They do not verify whether specific products meet given specifications; this is the purpose of the product audit, but the examination has to establish whether the system will cause products and service to be produced which will meet given specifications and prevent nonconforming products and service from being supplied[14]. In this regard the quality system assessment is more than a confirmation that procedures are being followed. Following a procedure will not guarantee that the results meet specified requirements since the user of the procedure will have to make judgements on the information obtained through the procedure. Procedures can only cause a course of action to be followed; they cannot predict the result. It is therefore necessary that in determining capability the results of using procedures are examined to establish that they are accurate.

---

[13]The scope statement in ISO 9001 declares that the standard is used where a supplier's capability to design and supply conforming products needs to be demonstrated.

[14]Also in the scope of ISO 9001, it is declared that the standard is aimed primarily at achieving customer satisfaction by preventing nonconformity at all stages from design through to servicing.

To fulfil these needs a quality system assessment would encompass:

- An audit of documentation to verify that the system will cause conforming product to be supplied and that adequate provisions are included to prevent nonconforming product being supplied

- An audit of records to establish accuracy

- An evaluation of performance to determine capability and effectiveness

- An assessment of results to decide whether or not registration to a prescribed standard should be recommended

It follows therefore that quality system audits or assessments comprise four elements, as illustrated in Figure 1-1. We shall use the term 'quality system assessments' when referring to the complete cycle of activities but 'audits' when referring to the documentation or implementation element.

Not all audits will warrant an evaluation element. Evaluation may be carried out by someone other than the auditor. The body commissioning the audit may in fact wish to evaluate the results or may contract the auditor to carry out this activity and a certificate may not always be awarded.

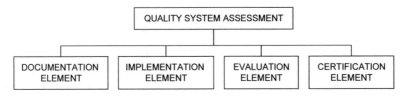

Figure 1-1 The elements of a quality system assessment

Documentation element

The documentation element comprises an examination of the documented quality system to verify that the requirements of the relevant ISO 9000 series standard and contracts etc. have been adequately addressed. 'Adequate' in this context means that the system will cause conforming product to be supplied and that provisions are included to prevent nonconforming product being supplied.

Implementation element

The implementation element comprises an examination of practices to establish that the provisions of the documented quality system are being properly implemented in so far as they address the requirements of the standard or contract. Many quality systems go beyond the minimum requirements of ISO 9000. The purpose of the implementation audit is to verify that the provisions relative to ISO 9000 have been implemented and not any other provisions such as occupational health and safety and environmental provisions unless covered by the contract. In examining the practices the auditor needs to establish:

- The situation as formally described and displayed (the manifest situation)

- The situation as it is assumed to be by the auditee (the assumed situation)

- The situation as revealed by systematic exploration and analysis (the extant situation)

- The situation as it would have to be to accord the requirements of the standard (the requisite situation)

The ideal situation is that in which the manifest, the assumed, the extant and the requisite are as closely as possible in line with each other[15]. Where they are not in-line, there may be justification for the auditee to take corrective action.

Evaluation element

The evaluation element comprises an assessment of the results of the previous documentation and implementation activities and the drawing of conclusions as to the effectiveness of the system. A system is effective if it provides the organization with the capability to supply specific products and services that meet customer needs and expectations.

Certification element

Certificates are not normally awarded on internal audits; however, some customers award high performing suppliers with certificates and all certification bodies award certificates to those that meet the success criteria. The certificate is awarded by the company that carried out the assessment and not the auditor. Auditors can only make

[15]Wilfred Brown, *Exploration in Management*, Pelican Books 1965

a recommendation to their organization that the company be considered for registration and a certification be granted.

Internal and external quality system assessments

When carrying out an internal quality system assessment, the purpose is to establish that the system is capable of achieving defined objectives and is being implemented. Whatever the organization has specified as its objectives, the system has to be verified as capable of meeting them. Whatever the system procedures require, their implementation has to be verified.

With an external quality system assessment the objective is different. External quality system assessments serve to obtain assurance that the system is capable of causing the supply of conforming products and services and preventing the supply of nonconforming products and services. Many external assessments are performed against ISO 9001. ISO 9001 is a model for quality assurance and therefore does not cover all the provisions needed for an organization to run its business. ISO 9001 contains the minimum requirements for obtaining an assurance of product or service quality. As such, an ISO 9001 audit will not test compliance with provisions of the organization's quality system that are beyond the standard. In complying with ISO 9001, organizations have to be capable of demonstrating that their system will assure product and service quality rather than that they have documented what they do and done what they documented. This is often a source of misunderstanding with respect to the difference between quality management and quality assurance. Organizations should design their quality systems to meet ISO 9004 and then select an appropriate standard for demonstrating to customers that their system will give an assurance of quality. In this regard the auditor only tests compliance with the requirements of the specified standard. One way of looking at it is to consider ISO 9004 as the design requirement for the quality system and ISO 9001 as the test specification. Not all provisions of a system designed to meet ISO 9004 are tested to satisfy ISO 9001. For example ISO 9004 addresses quality costs and product liability. These aspects are not tested in an ISO 9001 audit unless they are invoked in contract or are specified as a means to meet a requirement of ISO 9001.

Organizations will often design their systems to meet ISO 9001, neglecting ISO 9004. In this respect, their systems will be minimal and only serve the need for assurance. Even so many of these systems go beyond the standard as they perceive that those who need the assurance of quality include upper management and, regardless of customer requirements, upper management need assurance that their policies are being applied consistently.

Auditors often neglect this subtle distinction between quality management and quality assurance. Not everything has to be demonstrated to obtain an assurance of product quality. This is why the determination of the audit scope is of prime importance (see Chapter 3). In a second party audit the standard is likely to include contract requirements which may well go beyond those specified in ISO 9001. For instance, in the automobile industry, Ford, General Motors and Chrysler have jointly produced a version of ISO 9001 called QS 9000. This standard goes well beyond what ISO 9001 requires and includes such provisions as Failure Modes Analysis, Statistical Process Control, Quality Improvement and compliance with government Health, Safety and Environmental legislation. (See also Appendix B.)

First party auditors may use ISO 9001 as the standard against which performance is judged. In these cases, the standard is often applied to all departments regardless of their direct impact on deliverable products and services.

Third party auditors should beware that ISO 9001 is limited in what it requires and for the certification to be of any value the criteria for registration have to be equitable. If one auditor chooses to regard nonconformities with provisions outside the scope of the standard as pertinent to the certification and another limits the findings to provisions within the scope of the standard then one firm could be refused certification even if it complied with all the requirements of ISO 9001. Another firm may be granted certification if it did not implement the provisions outside the scope of the standard. We must have a level playing field if certification is to mean anything and be recognized across continents. It is similar to sitting a school examination. You are only tested on certain subjects and if you pass then, irrespective of your total knowledge, you have reached the same standard as everyone else who passed. There may be some with far greater knowledge than yourself but they receive the same certificate as you do. Similarly with ISO 9001, all receive the same certificate no matter how far the organization's quality system goes beyond the scope of the standard. If an organization wishes to gain recognition for the efforts it has put in beyond ISO 9001 then it can seek certification to Environmental Standards, Safety Standards or apply for a National Quality Award.

The audit process

A model of the generic quality system audit process is illustrated in Figure 1-2. Process models for internal, second and third party quality system assessments are illustrated in Figures 1-3, 1-4 and 1-5 respectively. The steps involved at each stage may differ depending on the type of audit and the audit objectives. Each stage is analysed further in subsequent chapters. The principles involved at each stage are as follows. These principles apply only to management audits, quality system audits and procedure audits and not to product and process audits.

The audit programme

- Internal audit programmes address certain parts of the business, either by department or by process whereas external audit programmes address organizations, contracts or projects.

- Audit programmes define the subject, location and date of the audit and may identify the name of the auditor or Lead Auditor.

- Audit programmes are duration limited either annually or for contract/project duration.

- Audit programmes are prepared by the Audit Organization rather than the auditor unless they are one and the same.

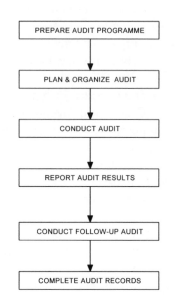

Figure 1-2 The generic audit process

Planning and organizing the audit

- Each audit in the audit programme should be defined by an audit brief prepared by the audit organization.

- The Lead Auditor is appointed by the audit organization and given the audit brief.

- The Lead Auditor is responsible for planning and organizing the audit.

- Contact with the company is through the Lead Auditor.

- The audit team should be selected on ability and expertise in the particular business sector.

- Pre-audit visits are needed to gather information for audit planning purposes.

- Preliminary assessments are carried out to determine the readiness of the organization for a formal assessment.

- Examination of quality system documentation is needed for audit planning purposes as well as for verifying compliance with standards.

- If the documentation audit is performed several months before the implementation audit, then the company may be requested to correct the documentation prior to the implementation audit.

- Audit schedules define which areas of the business will be audited by which auditors at what times.

- Audit check lists indicate what auditors will do when in each area and are guides to help auditors cover the scope and depth required.

- Auditors should be briefed by the Lead Auditor before they proceed to the site.

Conducting the audit

- The quality system documentation is submitted and examined before proceeding with the implementation audit and can be carried out off-site or on-site.

- The implementation audit commences with an Opening Meeting between the auditors and the management to agree the purpose, scope and conduct of the audit.

- The implementation audit takes place through interviews with staff performing activities which affect product or service quality.

- Objective evidence is gathered on representative samples of activities being carried out and on the documentation used both to perform the activities and to record the results of the activities.

- Objective evidence is confirmed with the organization's representatives during the audit.

- The objective evidence is analysed and decisions made as to whether the sampled documentation and/or activities conform to the specified requirements.

- Deviations from requirements are assessed to determine their effect on product or service quality and the effectiveness of the quality system and conclusions made as to whether the audit objectives have been achieved.

Reporting the audit

- The findings of the auditor are reported verbally during the audit.

- The results of the audit are documented in the form of a report detailing the conformities and nonconformities with the prescribed standard together with other observations which signify opportunities for improvement.

- The results and conclusions of the audit are reported at a Closing Meeting between the auditors and the management.

- Corrective action proposals offered by the organization are agreed and a schedule established for evaluating their effectiveness.

Following up the audit

- Corrective action proposals are submitted to the auditor for evaluation and acceptance.

- The auditor evaluates the proposed action and signifies acceptance or rejection.

- If agreed, the proposed actions are implemented by the organization. If the proposed actions are not agreed, revised proposals are presented until agreed.

- Implementation of the agreed actions are checked by the auditor either through correspondence or a repeat site audit to verify that the original nonconformity has been resolved effectively.

Completing audit records

- Completion of corrective actions is tracked by the auditor and liaison with the organization maintained.

- When all actions have been satisfactorily completed the audit is closed and the report updated and filed.

- Where appropriate the relevant certificates are then issued.

- Dates are established for continuing surveillance audits based on the results of the previous audit and the organization's response.

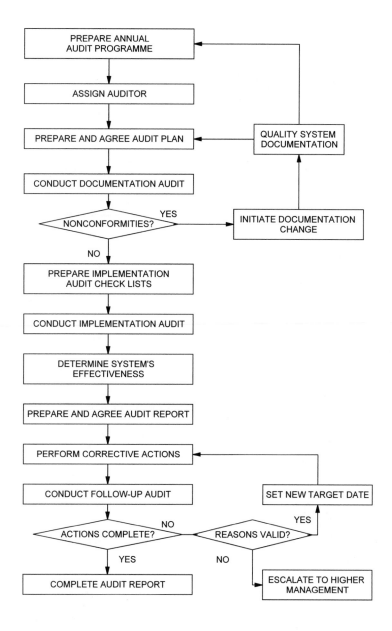

Figure 1-3 Internal quality system assessment process

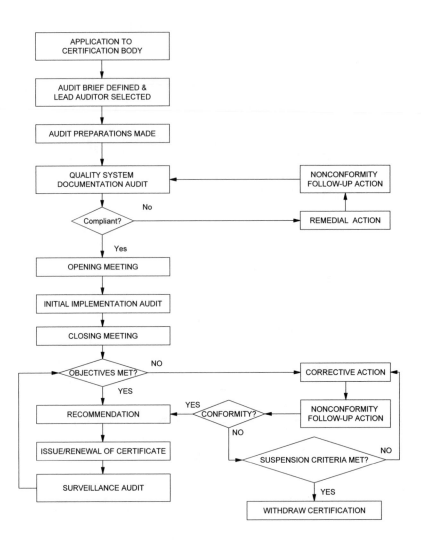

Figure 1-4 External (third party) quality system assessment

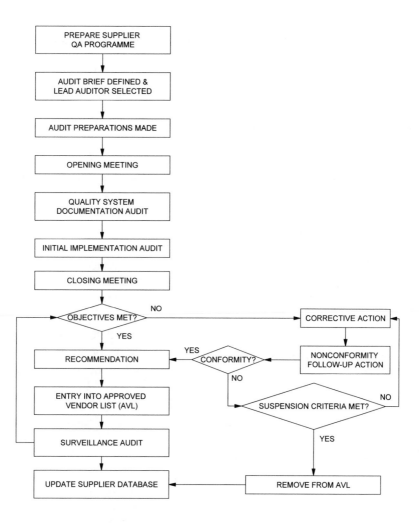

Figure 1-5 External (second party) quality system assessment

Summary

This chapter has explored the principles of quality auditing and the definitions used in this field.

◻ Audits establish facts not faults.

◻ All audits examine performance against given standards to determine compliance with these standards.

◻ There are many types of audit, each relating to the objectives to be achieved.

◻ There are also several categories of audit, known as first, second and third party audits.

◻ Terminology in the auditing field is fluid. Many concepts are known by different labels so it is necessary to understand what the other person is talking about before jumping to conclusions.

◻ Quality system assessments are not simply audits to determine conformance to procedure but audits to determine the capability of an organization to supply conforming products and prevent the supply of nonconforming products.

◻ A complete quality system assessment consists of four elements: documentation, implementation, evaluation and certification, which collectively represent an assessment of an organization's performance.

◻ The audit process is the same regardless of the type or category of audit and consists of a cycle of six stages: audit programmes, audit planning and organizing, audit conduct, audit reporting, audit follow-up and audit completion – after which the cycle is repeated.

Chapter 2

Audit management

In this chapter, the aspects covered by ISO 10011 on audit programme management and other relevant matters are addressed in detail. As with all types of management, auditing is no different in that it needs to be planned, organized and controlled if it is to be effectively performed. The effective management of audits depends on several key factors:

- Clearly stated audit objectives which are measurable

- The development of an audit programme to meet the audit objectives and to co-ordinate audit activities

- The acquisition and organization of competent auditors to execute the audit programme

- The development of effective auditing standards to measure auditor performance

- The development of effective procedures to execute audits

- The evaluation of the effectiveness of the audit programme and the auditors

- Improvement in performance should the audit objectives not be achieved or standards and procedures not followed.

Audit objectives

The objectives of particular audits need to be stated in the audit brief (see Chapter 3). All audits, however, should accomplish certain common objectives so that there is a means by which the audit organization may measure the quality of its service to the company and to society:

- To perform audits in compliance with defined standards within agreed timescales and budget

- To render a service which satisfies customer needs and expectations

- To preserve the confidentiality of all information obtained from all parties

- To achieve consistency among auditors and audit findings

The audit programme

Audit programmes should be prepared by the organization charged with carrying out audits. It should identify the type of audit, the dates on which the audits are to be conducted, the standard to be verified and the auditors responsible for performing them. Such programmes serve to co-ordinate the auditing activities of an organization so that work is planned to make effective utilization of available resources. There are two types of audit programme: one concerned with a single organization and the other with multiple organizations.

Single organization audit programmes

Single organization audit programmes should cover all aspects of the quality system in all areas in which it is employed. The coverage of the audit programme should be designed so that it obtains sufficient confidence in operations to be able to declare that the system is effective.

Multiple organization audit programmes

With multiple organization audit programmes they will show against each organization (the company) when initial, surveillance and repeat audits are due. Supplementary programmes may be necessary to show for each organization the elements of the quality system to be verified.

Audit standards

All audits should be conducted against a standard for the performance being measured. Examinations without such a standard are surveys and not audits. ISO 9000 is but one series of standards used for quality system audits. Audits can also be conducted against contracts, project plans, specifications, in fact any document with which the organization has declared it will comply. Industry sectors such as the automobile, medical, marine and aerospace sectors are tailoring ISO 9000 standards to suit their particular needs. They are adding requirements with which contractors have to comply to tender for contracts and, in spite of the aim of the ISO technical committee TC/176 which is responsible for the ISO 9000 series, the proliferation of quality system standards is on the increase.

Audit frequency

With first party audits, ISO 9001 requires that they be scheduled on the basis of status and importance of the activities being audited. Although applicable to internal audits, the concept is equally applicable to the scheduling of second and third party audits.

Status of activities

Status has three meanings in this context: the first to do with the relative position of the activity in the scheme of things, the second to do with the maturity of the activities and the third to do with the performance of activities. There is little point in conducting in-depth audits on activities which add least value. There is also little point auditing activities which have only just commenced. You need objective evidence of compliance and that may take some time to be generated. Where the results of previous audits have revealed a higher than average performance in an area (such as zero nonconformities on more than two occasions), the frequency of audits may be reduced. However, where the results indicate a lower than average performance (such as a much higher than average number of nonconformities), the frequency of audits should be increased.

Importance of activities

On the importance of the activity, you have to establish to whom is it important. To the customer, the managing director or society. You also need to establish the importance of the activity upon the effect of non-compliance with the planned arrangements. Importance also applies to what may appear minor decisions in the planning or design phase but if the decisions are incorrect they could result in major problems downstream. Such decisions could be the decision to issue a project plan that is incomplete or to release a design into production before completion of design verification, with the attitude that there is plenty of time as it can be completed later. Audits should verify that the appropriate controls are in place to detect such errors before it is too late.

All requirements of the standard should be verified in all areas at least once every one to three years. The status and importance of the activities will determine whether the audit is once a month, once a year or left for three years.

With third party audits following the initial audit, surveillance may confirm that individual elements of the system remain effective such that over a given period, all elements are reassessed. Alternatively, a complete audit of the quality system can be carried out periodically with surveillance visits being limited to checking the risk areas (see also Chapter 6).

An audit of one procedure or requirement in the standard in one area will not be conclusive evidence of compliance if the same procedures and requirements are also applicable to other areas. Where operations are under different managers but performing similar functions you cannot rely on the evidence from one area as management style, commitment and priorities will differ. In order to ensure that the audit programme is comprehensive a matrix may be needed to show what policies, procedures, standards etc. apply to which areas of the organization. One audit per year covering 10% of the quality system in 10% of the organization is hardly comprehensive. However, there are cases where such an approach is valid. If sufficient confidence has been acquired after conducting a comprehensive series of audits over some time, the audit programme can be adjusted so that it targets only those areas where change is most likely, auditing other areas less frequently.

Procedures will contain many provisions not all of which may be susceptible to verification at the time of the audit. This may be either due to time constraints or work for which the provisions do apply not being scheduled. It is therefore necessary to record which aspects have or have not been audited and engineer the programme so that over a one to three year cycle all provisions of the quality system and all requirements of the standard are audited in all areas at least once.

Organization of auditors

The organization which carries out audits should be equipped with the resources needed to discharge their obligations. This includes the acquisition of competent auditors and administrators.

The role of auditors

The audit team

When an audit is carried out by a single auditor this auditor assumes all the responsibilities and authority for planning, conducting and reporting the audit. When an audit requires more than one auditor then someone has to lead the audit and this person is known as the Lead Auditor. With a team of two or more auditors the roles of Lead Auditor and Auditor differ. The team may also include specialists who provide particular expertise, trainees, and observers who may be auditing the auditors or who have some other purpose which is acceptable to both the auditee and the Lead Auditor.

Lead Auditor role

The role of the Lead Auditor is to manage the audit in terms of its planning, organization and control. Depending on the size of the audit team the proportion of time spent auditing and managing will vary. The responsibilities and authority of the Lead Auditor are as follows.

Lead Auditor responsibilities

Planning

- Ensuring that audit brief is within the range of his/her capability before acceptance
- Gathering sufficient information on the company's products/services and particular industry sector for planning the audit
- Security of all documents obtained from the company
- Establishing how the company conducts its business
- Ensuring the documented quality system is compliant with the standard before commencing the implementation audit
- Preparing an Audit Schedule which is acceptable to the company
- Ensuring that the audit does not commence until adequate preparation has been made

Organizing

- Determining audit duration and team size commensurate with audit objective, schedule and financial limitations
- Selecting auditors who are competent for the task and are acceptable to the company
- Ensuring that auditors are fully briefed before commencing the audit
- Ensuring adequate health and safety protection is provided for the audit team during the audit
- Ensuring the protection of company interests
- Ensuring access for audit team to visit the specified location

- Ensuring that the audit team arrive at the right place at the right time

Controlling

- Conducting the Opening and Closing Meeting
- Establishing and maintaining cordial relations with the company
- Ensuring the audit is carried out according to the agreed plan
- Speedily resolving problems in conducting the audit
- Reporting the results of the audit to the company effectively
- Drawing accurate and valid conclusions from the results of the audit
- Submitting a documented audit report to the company

Lead Auditor authority

Planning

- Determining the date and duration of the audit
- Determining the range and depth of the audit

Organizing

- Selecting the audit team
- Allocating audit tasks
- Liaising with the company

Controlling

- Determining conduct and duration of Opening and Closing Meeting
- Entering any area in the business in which activities governed by the quality system are carried out.
- Terminating the audit
- Changing the plan to meet objective
- Deciding whether a company be recommended for certification or approval

Auditors' role

The role of the auditor is to plan, conduct and report the audit to the satisfaction of the Lead Auditor. Auditors will be assigned areas of the business to audit and be given a brief which will constrain the scope of the audit.

Auditor responsibilities

- Stay within the audit brief
- Determine the aspects to be verified in a given area to meet the audit objective
- Preparing audit check lists that probe the company's processes to a depth necessary to achieve the audit objective
- Establishing that the relevant parts of the quality system documentation meet the requirements of the standard
- Determine the extent to which the company's quality system is being implemented
- Collecting and documenting objective evidence without jeopardizing company relationships
- Accurately recording and clearly reporting the findings of the audit
- Safeguarding company documentation
- Notifying the Lead Auditor of any matters that may jeopardize the success of the audit
- Verifying the effectiveness of corrective actions
- Supporting the Lead Auditor
- Adhering to the prescribed code of conduct for auditors

Auditor's authority

- Entering any area in the business in which the activities carried out are within the scope of the audit
- Interviewing any company employee whose activities fall within the scope of the audit
- Selecting samples of work to verify compliance with prescribed requirements
- Requiring visibility of objective evidence to substantiate actions and decisions taken which are within the scope of the audit

Qualifications of auditors

Auditors should be qualified to perform quality system audits on the basis of appropriate education, personality, training and experience. However, audit technique, good communication skills and the right personality are far more important than formal academic qualifications providing the person has the ability to deal with masses of data and draw valid conclusions.

Education of auditors

Auditors should have a good general education and be literate, numerate and fluent in
their mother tongue. They should have demonstrated through examination the ability to
absorb data, analyse complex situations, discover true relationships between various
forms of data and draw logical conclusions. In

AUDITOR QUALIFICATIONS
• Education
• Personal characteristics
• Training
• Experience

short they should be qualified to degree level or equivalent.

Auditors should also either have academic qualifications in the particular business
sector in which they will conduct audits or have demonstrated the ability to research
and learn about such business sectors. In judging the level of academic attainment
auditors should in general have academic qualifications similar to the people they will
be interviewing in order that the interview can be conducted on at least equal intellect.
A person who failed the school leaving exam and then occupied a low position in
industry may not command the respect of the CEO.

Personal characteristics of auditors

Not everyone is suited to being an auditor just
as everyone is not be suited to being a lawyer,
a medical practitioner, an engineer etc. Quality
system auditors also differ from financial
auditors in that the standards used and objective evidence being sought are less precise and
require interpretation to the particular business sector.

Quality system auditors should be:

- Inquisitive so as to seek out information
  naturally

- Systematic so as to search for information
  in a logical and progressive manner

POSITIVE AUDITOR TRAITS	
• Systematic	• Inquisitive
• Analytical	• Assertive
• Mature	• Patient
• Flexible	• Honest
• Observant	• Direct
• Objective	• Perceptive
• Sensitive	• Friendly
• Tenacious	• Confident
• Alert	• Persistent

- Assertive so as to obtain information fairly and effectively

- Analytical so as to derive proper conclusions from collected facts

- Patient so as to tease information out of cautious, passive or aggressive people

- Mature so as to act responsibly and avoid being petty and childish in behaviour

- Honest so as to obtain information without recourse to trickery or deception

- Flexible so as to consider novel solutions to requirements without prejudice

- Direct so as to avoid wasting time and confusing the auditee

- Observant so as to discover important findings

- Perceptive so as to realize the implications of one's own actions and those of others

- Objective so as to maintain a focus on the audit purpose, arrive at demonstrable results and avoid nit picking

- Friendly so as to create a co-operative relationship with the interviewee

- Sensitive so as to react effectively to stressful situations, national conventions and customs

- Confident so as to ask searching questions and remain committed to conclusions despite pressure to change that is not based on objective evidence

- Tenacious so as to seek out objective evidence to support the company's verbal statements

- Persistent so as to follow the productive trails to the end

- Alert so as to recognize connections between what is being said and what is documented

Quality system auditors should not be:

- Aggressive such that interviewees get angry or upset

- Confrontational such that interviewee argues

- Passive such that interviewee controls the audit

- Dishonest such that information is revealed by trickery or deception

- Disorganized such that they can't find what they are looking for or search at random

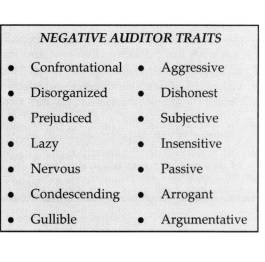

NEGATIVE AUDITOR TRAITS	
Confrontational	Aggressive
Disorganized	Dishonest
Prejudiced	Subjective
Lazy	Insensitive
Nervous	Passive
Condescending	Arrogant
Gullible	Argumentative

- Subjective such that their findings are vague and imprecise

- Prejudiced such that solutions are rejected that do not equate with the norm

- Insensitive such that the feelings of the interviewee are overlooked

A Tale About Check Lists

The audit team had been given a standard check list that was based on the requirements of ISO 9001 and were having difficulty determining how they could use this check list in the areas that had been assigned to them.

The Leader, Mustafa, turned to the flip chart and proceeded to draw a table showing where the elements of the standard matched the areas to be audited by each team member. Apart from Mustafa, the other auditors were fairly new to auditing. Although they had passed the approved Lead Auditor Course, they had had no real auditing experience to date.

The documentation audit had generated a number of queries that the team wished to take up with the company when on site so when Mustafa handed out copies of the appropriate check list to his team, they added a few additional questions to be asked.

The audit went well, or so Mustafa thought. However, at the Closing Meeting, the Quality Manager took Mustafa aside and told him about the conduct of the other auditors. He inquired why they were asking questions directly from their check lists, pointing out that in many instances the auditees could not relate the questions to what they did. The auditors seemed to dart from one thing to another – one moment on quality policy and the next on document control and no questions about the process.

Mustafa, defending his auditors, explained that they have to address all requirements of the standard, therefore all the questions were directly derived from the standard. The Quality Manager commented that whilst he had no doubt that the questions related to the standard, his impression was that the auditors made little attempt to understand how work was conducted in his company. He suggested that on future occasions the auditors follow trails through the process. There had been several instances where the auditor did ask pertinent questions but appeared to be adhering rigidly to the check list instead of exploring the answers that they were given.

The moral of this story is:

If you need a route map, don't use ISO 9001 as your guide!

Training of auditors

All quality system auditors should undergo formal training and retraining so as to equip them with the skills and knowledge needed for the job. It is therefore essential that the audit organization identify training needs continually and provide for the training of all its auditors in order to ensure they are competent to perform the tasks assigned to them.

ISO 10011 provides guidelines covering the education, training, personal attributes and experience requirements for quality system auditors. Training organizations wishing to run IRCA approved auditor training courses have to comply with specified Registration Criteria for such courses and are assessed periodically by the IRCA.

The approved auditor training courses cover auditing techniques, knowledge of ISO 9000 and quality management terminology but do not in general address other skills which auditors should possess to be effective auditors. The following list of topics covers a broad range of skills in which auditors should have either formal training or practical experience:

- The technologies that apply in the areas they will subsequently be auditing
- Quality system design, development and implementation
- Audit team management
- The interpretation of the standards used for auditing
- Planning, conducting and reporting audits
- Interpersonal skills
- Assertiveness skills
- Oral and written communication skills
- Personnel management skills
- Report writing skills
- Time management skills

Auditors should also maintain their competence by updating their knowledge of the standards and procedures employed and the business sectors in which they operate.

Experience of auditors

Auditors need experience both in conducting audits and in the business sector in which they will carry out their audits. If there is a gap of several months between audit training and actual practice, the knowledge and skills developed will quickly decline.

Therefore following audit training, auditors should carry out several audits under the supervision of an experienced auditor in order to hone their skills. If auditors are permitted to go straight from formal training into conducting actual audits without supervision there is a distinct possibility that they will not develop effectively and may well develop bad habits.

Although auditors who have several years experience of conducting audits may be able to carry out audits in business sectors in which they have had no practical experience, they may have developed knowledge which is inappropriate for certain businesses. What may be the norm for one sector may not be the norm for other sectors. The experience gained from working in a particular business sector equips the auditor with a knowledge and capability that will help him or her:

- Understand relationships between various work groups in a business
- Understand the processes through which product or information must pass to convert inputs into outputs
- Identify the risks, the important factors and the unimportant factors
- Understand what needs to be in place for operations to be under control
- Focus in on the key issues more quickly
- Make effective use of time
- Build a rapport with the interviewee
- Communicate effectively with management and staff
- Reach conclusions more quickly

However, there are often occasions when no auditors are available with experience in the particular business sector and in such cases it is reasonable to deploy qualified auditors having no relevant experience in the particular business sector providing they:

- Have a broad experience and an open mind
- Carry out research of the business sector to learn about the key features of such a business
- Are able to consult specialists and employ them in the audit team if necessary
- Are able to carry out a pre-audit visit to plan the audit and obtain the necessary information

Such a practice is called cross-training and provides a means for auditors with practical experience in one industry sector to gain experience in other sectors without having been employed in such sectors. If the rules were such that an auditor had to be employed in a particular industry to qualify to audit that industry, there would be a shortage of auditors for many industries. Sayle[1] condemns the school of thought that says that if one is acquainted with the principles of auditing, one can audit any activity and how right he is to do so. Even the auditor subject to cross-training needs to conduct several audits as a member of a team in the new industry sector before being deemed competent to lead an audit in such a sector. But without such techniques as cross-training the demand for auditors could not be satisfied.

In the UK there are several sector schemes – software, pharmaceutical and aerospace – where the industry sectors have declared that they require auditors qualified in their sector. The rules of such schemes prohibit cross-training since the auditor has to have recent experience in the industry to qualify. This presents a problem in that for anyone whose full-time job is auditing, the recency of industry experience will lapse and after three years this person will fail to be re-registered. It would seem that the schemes will only survive if there are sufficient people whose job involves both auditing and practical experience in the industry sector. As certification bodies prefer to employ auditors on permanent staff, the scheme will have to be resourced by freelance auditors and consultants.

In addition, auditors with experience in those activities of a business that deal with the determination or evaluation of results or the correction of failure are particularly valuable. Such experience provides investigatory skills and an awareness of the effect that failure to meet requirements has on product or service quality. Typical departments of a business in which such experience may be obtained are quality assurance, design verification, inspection, testing, customer service, customer support, installation, servicing and maintenance, and customer relations.

Selection of auditors

Auditors should be selected on the basis of appropriate education, training, personality and practical experience both as an auditor and in particular business sectors. For this reason it is unlikely that a person straight from school would have the necessary qualifications. Personnel without auditing experience can be trained if it is felt that they meet the other criteria. Personnel without the necessary intellectual capacity and personality cannot be trained, although it is important to distinguish between personality traits and behavioural weakness since the latter can be corrected. People can develop interpersonal skills but cannot change their personality. Also some apparent

---

[1] *Management Audits* by A J Sayle 1988

weaknesses may be due to a lack of self confidence which is easily improved through experience.

Auditors for particular assignments should be selected on the basis of the knowledge and experience they have of:

- The type of business sector: i.e. manufacturing industry, process industry, software industry, service industry and the associated statutory and regulatory requirements

- The type of market: i.e. domestic, consumer, contracting, government, defence etc. and the typical customer expectations and needs

- Type of product: i.e. engineering, construction, design and the typical technical problems encountered

- Type of service: i.e. leisure, finance, health, transportation etc. and the typical technical problems encountered

- The required language skills

- The size of the business; i.e. auditors experienced in large companies may not be suitable for auditing small companies and vice versa

- The national customs and conventions

The Lead Auditor should be selected on his/her ability to manage a team of auditors, communicate with top management and secure their confidence.

Auditing standards

Auditing standards are the standards to be achieved by auditors rather than the standards against which audits are to be conducted. Auditing standards cover the code of conduct and performance and address the following as appropriate.

Code of conduct

Auditors registered with the IRCA accept compliance with a code of conduct. The following list summarizes and extends these requirements where indicated:

- Operate in a trustworthy and ethical manner without prejudice or bias

- Disclose any conflict of interest prior to accepting an audit assignment. A conflict may arise when the auditor has shares or stock in the company to be audited, has been an owner or employee or has been the subject of an enquiry in a previous audit.

- Not accept bribes

- Treat all information provided by the auditee as confidential

- Not undertake assignments for which he/she is not competent to perform or available to complete (additional)

- Not charge the company for work that has not been carried out (additional)

- Not withhold evidence likely to affect the audit conclusions and recommendations (additional)

The three additional requirements are related to the auditor's professionalism which is not covered by the IRCA code of practice. The IRCA requires auditors to be trustworthy and unbiased which is only part of what being professional is about.

Performance standard

In planning, conducting and reporting audits, auditors should:

- Not commence an audit without adequate preparation

- Not carry out an audit without prior notification to the auditee unless otherwise agreed

- Not go beyond the agreed scope of the audit

- Take account of the results of previous audits if available

- Verify the effectiveness of any outstanding corrective actions from previous audits before leaving the site

- Verify compliance with all requirements of the standard on a sample of operations

- Be consistent in the interpretation of the standards and judgement of compliance

- Report all findings to the auditee including good points and observations before leaving the site of the audit

- Not disclose results that cannot be substantiated by objective evidence

- Withdraw nonconformity statements when the company proves the requirements are not applicable or when further objective evidence demonstrating conformity is presented

- Provide sufficient information in the audit report to substantiate the conclusions and recommendations

- Complete the audit within the agreed timescales

- Advise the auditee of any difficulties in conducting the audit within the agreed timescales that are beyond the auditor's control

Audit procedures

Procedures should be established, documented, implemented and maintained to create consistency and predictability in the audit process. These procedures should prescribe how quality system audits are to be planned, conducted, reported and completed and should address the following subjects as a minimum:

- Preparing and maintaining the audit programme
- Selection of auditors
- Audit preparation
- Audit planning
- Arrangements for conducting out-of-country audits
- Conducting the audit
- Recording observations
- Classification of nonconformities
- Reporting audit findings
- Determining audit conclusions and recommendations
- Evaluation of corrective action proposals
- Confirming the effectiveness of corrective actions

- Quality system surveillance

- Dealing with appeals

- Maintenance of audit records

- Maintaining confidentiality

- The forms for planning the audit

- The forms for recording the observations, nonconformities and corrective actions

Audit evaluation

There are three aspects of audits which require evaluation; the effectiveness of the audit programme, the effectiveness of the audit procedures and the performance of the auditor.

Audit programme effectiveness

The effectiveness of the audit programme can be evaluated by assessing:

- The frequency of audits from the results to detect whether frequency should be changed

- The resourcing of the programme to detect if adequate resources had been provided

- The coverage of the programme to detect whether all obligations have been honoured

- Work hours of auditors to detect auditor fatigue

Audit procedure effectiveness

The effectiveness of the auditing procedures can be evaluated by assessing:

- The degree of variation in auditing methods and results, since procedures should minimize variation

- The degree of flexibility, since procedures should not unduly constrain the audit

- The extent to which informal methods are employed by individual auditors to provide consistency

- Company feedback on administrative, planning and reporting matters

Auditor performance evaluation

The performance of auditors should be continually evaluated to establish opportunities for improvement. This is also covered in ISO 10011. The evaluation of auditor performance is often difficult as their performance cannot be assessed without observation. Invariably, the presence of someone monitoring an auditor is likely to have a positive or negative effect on the auditor's performance. There are several sources of information on auditor performance:

- Observation by more senior auditors either leading the team or independent of the team

- Review of audit reports

- Company feedback

- Audit discussion groups

- Simulated audits under controlled conditions within the audit organization

- Alternating between auditors for specific companies

- Periodic written examination and role play

The most reliable source of auditor performance data will be collected by his/her Lead Auditor as they will have the opportunity to observe the auditor in action for some of the time. However, since auditors operate alone and not in pairs, it may take some time before deficiencies are revealed. Lead Auditors are able to assess reporting skills but not interview skills although audit findings can reveal whether an auditor probed deeply enough to obtain the reported evidence. Lead Auditors should be required to carry out an appraisal of their teams' performance after each audit and this information can be analysed periodically by the audit organization to assess overall performance.

Improvement

Having evaluated the effectiveness of the audit programme, the audit procedures and auditor performance, corrective action should be taken to prevent the recurrence of unsatisfactory performance. Such action may require:

- Retraining of auditors

- Revision of procedures

- Re-organization of audit teams

- Acquisition of additional auditors or special resources

Even where corrective action is not appropriate, trends should be monitored and action taken to prevent auditor fatigue and customer complaint.

Summary

This chapter addressed audit management and its importance in setting audit objectives, organizing the audit, controlling and improving auditor performance. The following are the key issues covered:

◻ Audit objectives establish criteria for measuring audit performance.

◻ Audit programmes serve to co-ordinate the auditing activities of an organization.

◻ ISO 9000 is not the only standard for quality systems.

◻ Scheduling external audits on the basis of status and importance is as applicable as it is to internal audits.

◻ The effectiveness of audits depends on securing competent auditors equipped with the appropriate education, experience, training and personal attributes.

◻ Lead Auditors are auditors who lead audits and take responsibility for their planning, organization and control.

◻ Auditors must support the Lead Auditor and stay within the audit brief.

◻ Auditors need experience in the industry they are auditing, knowledge of the standard and the business processes and most of all, good communication skills.

◻ Auditors must be assertive and not aggressive or passive.

◻ All auditors should adhere to a code of conduct so that professional standards are maintained.

◻ Documented procedures are as essential for auditors as they are to the operations they audit.

◻ The effectiveness of auditing needs to be monitored continually so that auditing standards are maintained.

◻ Continual improvement in auditor performance is essential to maintain company confidence.

Chapter 3

Planning audits

The planning process

The audit programme identifies what audits are to be conducted in a given time period for the audit organization. Audit planning deals with a particular audit in the programme. The planning process commences with preparation of the audit brief and ends when the auditors arrive on site to commence the audit. This process is illustrated in Figure 3.1.

The audit brief

No audit should be carried out without an Audit Brief. The Brief is the audit requirement and without one the auditor may waste time and cause untold harm to relationships between the auditor's organization and the auditee. The Brief should define the following.

> **THE AUDIT BRIEF**
>
> - Purpose of audit
> - Organization identity
> - Scope of audit
> - Constraints
> - Contacts

The purpose of the audit

The purpose of the audit could be any one of the following:

- Compliance with contractual requirements
- Acceptance of products from a subcontractor
- Determination of capability to produce certain products

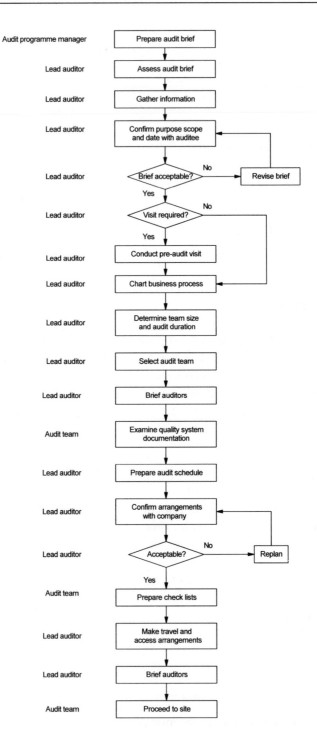

Figure 3-1 Audit planning process

- Entry onto List of Preferred Suppliers

- Certification to an international quality system standard

- Surveillance of the registered organization

Identity of organization

This would be the name, postal address, telephone and fax numbers.

Scope of the audit

The scope of the audit is the processes to be covered by the audit or excluded from the audit; if a Certification Audit, the scope of registration addressing the design, development, production, installation or servicing of certain named products and services as appropriate.

Constraints

Any constraints on dates and duration, number of auditors, elements of quality system not controlled at particular location etc.

Contacts

The names of two contacts who will liaise with the Lead Auditor are needed but it is not essential.

Preparation check list

There are a number of things that have to be carried out or obtained prior to carrying out the audit. The list that follows may help guide this process.

> **PREPARATION CHECK LIST**
>
> - Information gathering
> - Planning tasks
> - Logistics
> - Checks with auditee
> - Checks prior to departure

Information gathering

- Accept audit brief and clarify any ambiguities.

- Arrange date for pre-audit visit if cost effective.

- Prepare pre-audit visit check list.

- Obtain and study books, articles and standards relating to the particular industry sector and its products and services.

- Obtain copies of previous audit reports if appropriate.

- Obtain copies of supplier performance data if available.
- Complete pre-audit checks by visit or other means.
- Register all documents obtained from pre-audit checks.

Planning tasks

- Examine quality system documentation.
- Prepare business operations chart.
- Determine audit duration.
- Determine audit team size.
- Select auditors.
- Brief auditors.
- Prepare Audit Schedule (for each site if necessary).
- Prepare check lists.

Logistics

- Security clearance if required
- Maps and directions
- Transport arrangements to audit location
- Overnight accommodation
- Local travel arrangements (hire car, host company car etc.)
- Entry visas and work permits

Confirm with auditee

- Acceptability of Audit Team members
- Appointments with key managers
- Meeting rooms for Opening/Closing Meetings
- Meeting rooms for Auditors
- Attendees at Opening and Closing Meetings
- Lunch arrangements
- Car parking
- Protective clothing

- Health and safety procedures and declarations required
- Dress code
- Security
- Selection, availability and role of Guides

Prior to departing

- Copy of relevant standards
- Copies of blank report forms
- Copies of the previous assessment report
- Agenda for Opening Meeting
- Proforma agenda for Closing Meeting
- Clipboards
- Security clearance
- Confidentiality declaration
- Attendance list completed for audit team
- Audit schedule
- Audit check lists
- Auditees documents to be returned

Pre-audit visit

It is often wise to visit the site where the audit is to be conducted in advance of the audit so that a realistic plan can be produced. This is particularly important if an initial audit is being planned rather than a surveillance audit when a telephone conversation or an exchange of correspondence may be adequate. A Pre-Audit Check List should cover the following, as appropriate.

PRE-AUDIT VISIT

- Business aspects
- Products and services
- Personnel
- Facilities
- Documentation
- Audit details

Business

- Type of industry sector such as service, manufacturing, process, software

- Scope of business

- Scope of quality system

- Turnover

- Age and development history

Products and services

- What products and services are supplied

- Types of contracts that are executed by the organization

- Key quality characteristics for the products and services

- National standards relating to the products and services with which the company declares compliance

- Typical control issues of particular industry, technology, processes etc.

- Expected norms for defect levels

- Processes for which capability needs to be assured

- Key failure modes critical to product or service performance

Personnel

- Identify key managers

- Availability of key managers for the duration of the audit

- Number of guides

- Availability of guides for the duration of the audit

- Explain role of guides and selection criteria for guides

- Working hours

- Staff numbers and distribution

- Need for interpreters

Facilities

- Obtain site plan

- Explore site to determine complexity and accessibility

- Number of sites and their contribution
- Protective clothing requirements, health and safety issues
- Meeting room availability
- Lunch arrangements
- Parking restrictions

Documentation

- Quality system documentation structure
- Obtain copies of relevant manuals
- Quality system procedures composition and complexity
- Obtain sample of procedures and instructions
- Organization chart
- Promotional literature

Audit details

- Scope of registration and pertinent standard
- Explain audit process
- Explain the company's responsibilities
- Explain the success criteria
- Audit dates
- Existing approvals
- Outline Audit Schedule
- Identify processes requiring specialist auditors
- Security
- Concerns

Clarifying the purpose and scope of the audit

The audit brief will have identified the purpose and scope of the audit but there are several things to clarify:

- Which standard is to apply

- Which processes are to be included
- Which products and services are to be covered
- Which locations are to be included

The scope statement will eventually be included in the directory of organizations of assessed capability and will therefore play a role in advertising the capability of the organization. If these scope statements portray an organization that has a capability it is not prepared to offer to customers then such statements could be misleading. For example, if an organization claims it has a design capability when, although designing its own products, is not prepared to offer design as a service to potential customers and will only offer products for sale of existing design, then inclusion of design in the scope statement could imply that it does offer design services. The solution to removing the confusion is in how the company is listed in the directory of registered organizations.

In making an application for assessment, a company would have to declare the products and services offered. These are then coded and in the directory of registered organizations the companies are listed in several ways: by name, by product or service and by sector. Companies are therefore able to choose whether to be listed for design, manufacture, service, installation, distribution etc. The product codes used in the UK directory are as follows and whilst providing a means of classification, they are likely to be misapplied. Many service organizations are listed against code S, when in fact they should be code M.

M	Manufacturer or provider of a service
S	Service repair or maintenance
C	Consultant or design services
I	Builder, installer or erector
D	Distributor or stockist
P	Product certification

In cases where a company performs design and wishes to be assessed against ISO 9001, but does not offer a design service to its customers, it can choose to omit design from the services offered. The rules regarding this situation are currently under review by the UK Accreditation Service (formally the NACCB). Some organizations have been known to include internal functions in the scope statements, such as purchasing, marketing etc., indicating that they have a purchasing and marketing capability. However, most organizations have this capability but do not include it in the scope statements as they do not offer such services to potential customers, they merely contribute to the productive functions of the business. Where companies wish to include all operations of the company they may apply for Company Wide Certification where each function is treated as a service to other functions. BSIQA offer a Company Wide Certification Scheme. However, until such time as the rules change, the principle is therefore to only include in the scope statement those products and services intended for sale to potential customers unless otherwise requested.

Which standard

Some companies may apply for ISO 9001 when ISO 9002 is more appropriate or vice versa. The simple maxim on which standard is this. If their customers specifically require them to carry out design or if they cannot satisfy their customer requirements without designing a product for them, then ISO 9001 applies. If they design products to meet what they predict are the customer needs and expectations and then supply proprietary product, ISO 9002 applies. If they can determine that they have met all the customer requirements by examination of the final product or service then ISO 9003 applies. If a company designs its own products and services it may request certification to ISO 9001 but is not compelled to do so although the new rules from UKAS may alter this situation. Where a company wishes the design functions to be audited, then the audit has to be against ISO 9001.

Which processes

The range of processes from which to select is governed by the scope of the standard and includes design, development, production, installation and servicing but may include other related processes which represent the core processes of the particular business. Examples include: maintenance, construction, repair, laundry, health care, manufacture and publishing. There are nearly as many processes as there are businesses. If a company manufactures product but does not design, install or service it then the process is simply 'manufacture.'

Which products and services

Quality system audits are performed to establish whether the system is capable of enabling the company to design, manufacture etc. certain products and services. They may not have the capability to design and manufacture other similar products. For example, a company that produces analogue communications systems may not have the capability to produce digital communications systems. A company that produces cotton textiles may not have the capability to produce nylon textiles. The particular products and services are therefore based on what the company is equipped to design and/or produce.

Which locations

The locations from which a company operates are also important since such locations may or may not be equipped with the capability to do the things that are done at other locations. A location is generally a separate postal address so that the addition of a new building on the same site would not extend the scope but opening an office or factory in another state, region or country would extend it.

Typical purpose and scope statements

The purpose and scope statements need to qualify the particular products/services, standard and location in terms which prescribe the boundary conditions for the audit. Merely stating that the audit is being performed to verify compliance with ISO 9001 or ISO 9002 is inadequate. Some sample purpose and scope statements are given in the box below.

SCOPE STATEMENTS

The purpose of the audit is to establish the extent to which the quality system of Byte Computing Inc. of Santa Clara meets the requirements of ISO 9001 for the design, development, production, installation and servicing of main frame computers and associated peripheral devices.

The purpose of the audit is to establish the extent to which the quality system of Salmon Services of Portsmouth meets the requirements of ISO 9002 for the maintenance of sea going vessels up to 200 tonnes.

The purpose of the audit is to establish the extent to which the quality system of Jackolux Ltd of Bolton meets the requirements of ISO 9001 for the design and production of domestic glassware.

Explaining the audit process

It is important that the company understands the audit process so that it does not bring any surprises. It is also useful to allay any misunderstandings to prevent any problems that may be anticipated. Some companies may take the view that their staff should not volunteer information, leaving the auditors to discover for themselves the objective evidence. This can have the effect of stifling interviews and resulting in stress for both the auditor and auditee.

In explaining the audit process, the auditor should emphasize that:

- Auditors will be seeking conformance to the requirements of the standard rather than nonconformity.

- They will interview members of staff responsible for work that affects product or service quality and will examine samples of work and associated documentation.

- They will record any findings and seek agreement with the guides.

- Findings will be classified as either observations or nonconformities and will be presented for agreement to the Management Representative at the end of each day.

- All the results of the audit will be reported at a Closing Meeting on the last day when a hand-written report will be provided and a date agreed for the submission of any corrective action proposals.

- The company will be informed of the recommendations that the team will make to the Certification Body before leaving the site.

The auditor should also explain the role of the guides that are needed and how they should be selected (see later).

Company's responsibilities

The company has certain responsibilities towards the audit so that it yields maximum benefits:

- To determine the scope of the audit

- To inform their staff of the purpose, scope and objectives of the audit and the manner in which it will be carried out

- To appoint guides or escorts for the audit team

- To ensure that key personnel are available as required for the duration of the audit

- To provide facilities, local transport, communication and photocopying services

- To provide access to all facilities, locations, documentation, equipment relevant to the scope of the audit

- To co-operate with the auditors in the discovery of objective evidence that serves the audit objectives

- To determine, submit and implement the corrective actions necessary to prevent the recurrence of the detected nonconformities

- To provide access to the auditors to verify the effectiveness of corrective actions

The success criteria

The success criteria are those factors which indicate that the quality system is effective. Both the audit team and the company need to agree on the success criteria, otherwise there will be dispute and argument. The audit team needs to employ common criteria to determine whether findings are sufficiently significant to adversely affect the outcome of the audit. Differing definitions of success or failure can jeopardize the credibility of the conclusions. The company needs to be in agreement also so that the audit is seen as being fair and of added value. However, views differ on success criteria. There are many options ranging from an isolated error to a failure in the effectiveness of the system but these can be grouped into two classes: Random Failures and System Failures. The examples given are for illustration purposes only and may not be appropriate in a particular situation.

Random Failures

These are failures which have a low probability of recurrence and which require only remedial action to eliminate. The root cause is human error. There are two levels:

1   One instance of a failure to meet a requirement of a document forming part of the quality system. This is a lapse in following one instruction in a procedure, a plan, a specification or a contract. The requirement does not have to be traceable to a requirement in the standard: e.g. out-of-tolerance test results not recorded in Red Ink as specified in procedure QP 013 when there is no apparent reason for it; drawing check box not signed but approved box signed; training records not in location specified in procedure QP 018 but found in another location.

2   A failure to meet a requirement of the standard in the implementation of the quality system where the documented system provides for compliance. This is a random failure to meet one requirement of a paragraph in the standard: e.g. calibration status not indicated on AVO Meter S/N 346139; supplier of item KL 89756 not on approved suppliers list; box containing 100 KL 89756 components not in designated storage area.

System Failures

These are failures which have a high probability of recurrence and for which corrective action can be specified to eliminate the cause and prevent recurrence (see Appendix C on definitions of corrective action and preventive action). There are five levels, each reflecting a greater degree of severity:

1   A failure to meet one requirement of the standard in the design of the quality system: e.g. no procedure for the identification of training needs; no provision

for inspection records to define the acceptance criteria; no evidence that design reviews have been planned.

2   A failure to meet all the requirements of a clause of the standard in one area of an organization where the requirements apply to more than one process. This is a failure to provide the necessary controls. (A clause is a subject of the standard identified by a number such as 4.10.1.) For example, no training provisions for service engineers; no system for controlling documents and data produced by the maintenance department; the responsibilities and authority of test software programmers have not been defined and documented.

3   A failure to meet all the requirements of a clause of the standard in all processes of an organization where the requirements apply to more than one process. This is a failure to provide the necessary controls throughout the organization: e.g. no training provisions in place; no preventive action procedures in place; no internal audit programme in place; no management reviews of the quality system conducted.

4   A failure of the quality system to cause product or service conformity or prevent product or service nonconformity with specified requirements which substitute customer requirements (i.e. requirements specified by the supplier): e.g. no mechanism for preventing the supply of nonconforming product to customers; quality plans do not address means to achieve durability specification for notebook computer model S486-33; no measures found to verify that materials used in furniture met flammability requirements.

5   A failure of the quality system to cause product or service conformity or prevent product or service nonconformity with the expectations and needs of customers: e.g. no system in place to meet customer reliability requirements; no system in place to meet durability claims for notebook computer model S486-33; no system in place to fulfil the customer's flammability expectations for new furniture.

Nonconformities

ISO 8402 defines a nonconformity as a failure of the quality system to fulfil specified requirements and it defines corrective action as the action needed to eliminate the cause of a nonconformity. It is impractical to have seven types of nonconformity so a simpler classification is needed in order to prioritize the corrective action. Many auditors do regard random failures as nonconformities that require corrective action. In fact many audit reports only contain evidence of random failure such as a calibration label missing, an unapproved document or the absence of a training record for a particular person. Random failures could be corrected on the spot so need not be reported in the audit report other than as observations. Meaningful corrective action cannot be taken for the occasional random failure. Random failures should not warrant a decision to refuse certification unless their recurrence is excessive, implying a system breakdown. In such cases it indicates a systematic problem in the organization. System failures, however, need to be eliminated at any level and can be classified as indicated in the box. (Examples of these are given in Chapter 5.)

> *NONCONFORMITY CLASSIFICATIONS*
>
> *MAJOR*
>
> The absence or total breakdown of the provisions required to cause product conformity or prevent product nonconformity with the expectations and needs of customers.
>
> *MINOR*
>
> Any failure to meet one or more requirements of the standard. (A requirement in this context is defined in Chapter 7.)

Whilst many auditors do regard departures from internal procedures as nonconformities worthy of reporting, the reason for not reporting all internal procedure deviations is that the external audit is being conducted against the requirements of that standard. It is essential that every nonconformity is traceable to a requirement of the standard. If there is no requirement in the standard, then there is no nonconformity. For example, a procedure may require test results indicating failures to be written in Red Ink with no apparent reason given. The standard does not require Red Ink to be used and, providing pass/fail criteria and the pass/fail decision is indicated, the organization is compliant with the standard regardless of the colour of the ink used. If it is found that the Red Ink is employed to facilitate a search for problems requiring corrective action as part of the response to clause 4.14.2, then the lack of such markings may result in no action being taken when action would obviously be necessary; in which case a request for corrective action may be appropriate. Auditors therefore have to use their judgement and search beyond the specific incident before reaching a conclusion. First party auditors would seek corrective action regardless of the use of Red Ink, whereas third party auditors are seeking evidence of conformity with the standard.

Examination of the quality system documentation

In order to plan an audit effectively the Lead Auditor should acquire copies of key quality system documents such as the Quality Manual, a list and sample of procedures and any other documents which describe the business, the way it is organized, its products and services if not included in the Quality Manual. Examination of these documents for planning purposes need not be as extensive as a complete check for compliance with the relevant standard although any significant nonconformities will be spotted. If elements of the standard have not been addressed then the Lead Auditor should discuss these with the company representative before proceeding further in order to avoid aborting the on-site audit. Depending on the complexity of the documentation, such nonconformities may be detected during the pre-audit visit. For planning purposes, things to look for are:

- Organizational relationships particularly internal and external interfaces

- Key processes and support services

- Distribution of work on site and off site

- Range of products and services offered

- Flow charts or procedural sequences which help to identify audit trails from process inputs to process outputs

- Requirements of the standard which are deemed not applicable

- Specific solutions to requirements of the standard

Any aspects that are unclear can either be clarified by contacting the company or by adding to the check lists for further investigation on site. Specific solutions which the organization has developed can also be added to the check list for verification on site. Examples are:

- Use of an approved vendor list as a means of selecting suppliers with no provision for abnormal situations

- Use of identified areas to denote inspection status with no provision for product marking

- Use of an electronic documentation system that shows no evidence of document approval

- In-line rectification of nonconformities with no system for identification, segregation etc.

The documentation audit is addressed in Chapter 4 on Conducting Audits since it is a part of quality system assessment even though it may be carried out before the Opening Meeting on site.

Identifying the business processes

Once the key documents have been obtained and the site visited, it should be possible to build a model of the business processes and identify the processes to audit.

It is neither practical nor possible to audit an organization against the standard element by element since no organization is formed in this way. The elements of the standard do not reflect either departments or processes. They are topics that departments and processes must address. One can audit by department but business is carried out across department boundaries rather that through one department after another. By auditing departmentally one will not necessarily check the interfaces and it is at the interfaces that communications often breakdown. By far the best method is to audit processes. However, before planning an audit the various processes of the particular business need to be determined and there are some simple techniques that can be used to do this:

Step 1   Identify the inputs and outputs of the business in terms of the customers, suppliers, products and services supplied. A generic model is illustrated in Figure 3-2.

Step 2   Identify the core processes which convert these inputs into outputs and join them by communication pathways. A generic model is illustrated in Figure 3-3.

Step 3   Identify the support processes upon which the core processes depend for their resources.

Step 4   Add the quality system management processes.

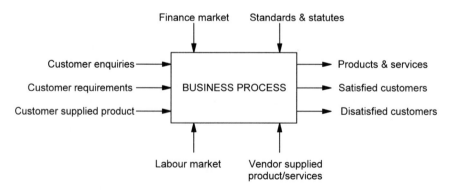

Figure 3-2 Level 1 Business process model

The business process model (Figure 3-2) shows the basic inputs and outputs of any business. In planning a specific audit the specific inputs and outputs should be identified. There may well be others.

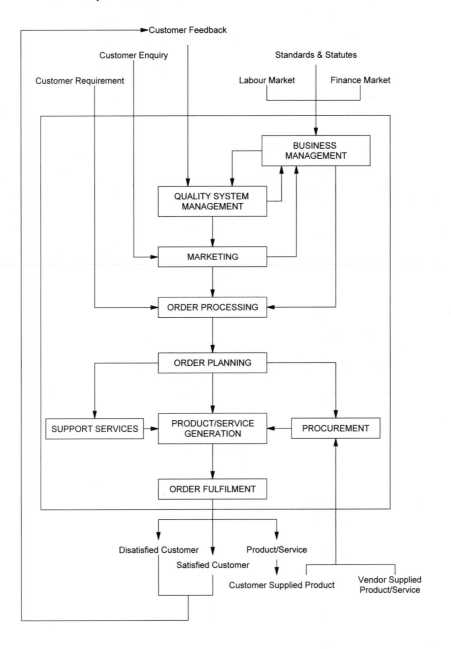

Figure 3-3 Level 2 Business management system model

In the business management system model (Figure 3-3) the processes consist of the following elements, which can be related to the clauses of ISO 9001 as indicated generally in Table 3-1 and in detail under each heading.

FUNCTION	4.1	4.2	4.3	4.4	4.5	4.6	4.7	4.8	4.9	4.10	4.11	4.12	4.13	4.14	4.15	4.16	4.17	4.18	4.19	4.20
Business management	X	X			X									X		X	X	X		
Quality system management	X	X			X									X		X	X	X		
Marketing	X	X	X		X											X		X		
Order processing	X	X	X		X											X		X		
Order planning	X	X		X	X				X							X		X		
Product/service generation	X	X		X	X		X	X	X	X	X	X	X	X	X	X		X	X	X
Procurement	X	X			X	X	X	X		X					X	X		X		X
Support services	X	X			X	X	X			X	X					X		X		X
Order fulfilment	X	X			X			X		X		X	X		X			X		X

Table 3-1 ISO 9001 element applicability matrix

Business management

- Purpose and mission. Although not specifically covered by the standard, the purpose and mission statements indicate organizational goals which are addressed by the standard (clause 4.1.1).

- Customer expectations and needs (clause 4.1.1)

- Business plans or organizational goals (clause 4.1.1)

- Quality policy, commitment (clause 4.1.1)

- Quality objectives (clause 4.1.1)

- Organization (clause 4.1.2.1)

- Resources (clause 4.1.2.3)

- Management review (clauses 4.1.3 and 4.14)

Quality systems management

- Quality system representative (clause 4.1.2.3)

- Quality system development (clauses 4.2 and 4.5)

- Quality system maintenance (clauses 4.5 and 4.14)

- Quality system audit (clause 4.17)

- Quality records (clause 4.16)

- Quality system performance (clause 4.1.2.1b)

Marketing

- Publicity. Although not specifically covered by the standard, publicity processes often set customer expectations and so needed to be controlled (clauses 4.1, 4.3, 4.5, 4.18).

- Tendering (clauses 4.1, 4.3, 4.5, 4.16, 4.18)

- Sales. Although not specifically covered by the standard, sales processes often attract customer enquiries which set customer expectations and result in orders being placed (clauses 4.1, 4.3, 4.5, 4.5, 4.16, 4.18).

Order processing

- Contract review (clauses 4.1, 4.3, 4.5, 4.16, 4.18)

Order planning

- Quality planning (clauses 4.1, 4.2.3, 4.5, 4.16, 4.18)

- Design planning (clauses 4.1, 4.4.2, 4.5, 4.16, 4.18)

- Production planning (clauses 4.1, 4.9, 4.5, 4.16, 4.18)

- Installation planning (clauses 4.1, 4.9, 4.5, 4.16, 4.18)

- Servicing planning (clauses 4.1, 4.9, 4.5, 4.16, 4.18)

Product/service generation

- Design (clauses 4.1, 4.4, 4.5, 4.14, 4.16, 4.18)

- Production (clauses 4.1, 4.5, 4.8, 4.9, 4.15, 4.14, 4.16, 4.18)

- Installation (clauses 4.1, 4.5, 4.8, 4.9, 4.14, 4.15, 4.16, 4.18)

- Servicing (clauses 4.1, 4.5, 4.9, 4.14, 4.15, 4.16, , 4.18, 4.19)

- Inspection and test (clauses 4.1, 4.5, 4.10.3, 4.10.4, 4.10.5, 4.12, 4.13, 4.14, 4.15, 4.16, 4.18, 4.20)

- Stores (clauses 4.1, 4.5, 4.7, 4.8, 4.15, 4.18)

Procurement

- Purchasing (clauses 4.1, 4.5, 4.6, 4.8, 4.16, 4.18)

- Receipt inspection (clauses 4.1, 4.5, 4.7, 4.8, 4.10.2, 4.10.5, 4.11, 4.13, 4.16, 4.18, 4.20)

Support equipment/services

- Inspection measuring and test equipment (clauses 4.1, 4.5, 4.6, 4.7, 4.11, 4.16, 4.18)

- Jigs tools and fixtures. Although not specifically covered by the standard, jigs, tools and fixtures are often used to form product characteristics which are incapable of economic measurement by normal inspection or test and are therefore devices for accepting product which are covered by the standard (clauses 4.1, 4.5, 4.6, 4.7, 4.11, 4.16, 4.18).

- Test software and hardware (clauses 4.1, 4.5, 4.6, 4.7, 4.11, 4.16, 4.18)

- Test laboratories (clause 4.1, 4.5, 4.10, 4.11, 4.16, 4.18, 4.20)

Order fulfilment

- Picking and packing (clauses 4.1, 4.5, 4.10.3, 4.10.4, 4.12, 4.13, 4.15.2, 4.15.4, 4.15.5, 4.18, 4.20)

- Dispatch/delivery (clauses 4.1, 4.5, 4.8, 4.15.2, 4.15.6, 4.18)

- Transportation (clauses 4.1, 4.5, 4.15.2, 4.15.6, 4.18)

In the above list it will be observed that several of the clauses apply to all processes. They are 4.1, 4.5 and 4.18. In other words, the requirements covering management responsibility, documentation and training apply to all processes except clauses 4.1.2.3 and 4.1.3, which are addressed separately under Business Management and Quality System Management.

Audit duration

Audit duration depends upon seven factors:

- The size of the organization: size of the site or sites, numbers employed in each processes

- The complexity of the organization: different technologies, contract/project/product types, market requirements, product application

- The complexity of the documented quality system and ease of understanding

- The scope of the audit: full range through design to servicing or limited; complete system audit or surveillance

- The number of auditors employed: Lead plus generalists and specialists

- The competence of the auditors: recent graduates or seasoned professionals

- The competence of the auditees: first audit experience or seasoned hosts

There are tables[1] which can be used to estimate audit duration but they are inexact. The periods which have to be estimated are:

Initial visit	0.5 to 1 day
Preparation	0.2 to 2 days
Quality system documentation audit	0.2 to 3 days
On-site audit	2 to 15 days
Corrective action evaluation	0.1 to 1 days
Follow-up	0 to 3 days

These figures place a complete audit in the range 3 days to 25 days but they are only a guide. This is not the elapsed time but the total number of auditor days. With one auditor, elapsed time would vary between 3 and 25 days. With a team of four auditors, the elapsed time would vary between 1.5 and 11.5 days. There is no magic formula, as every organization is different, which is why the initial site visit is so important. If an initial visit is not possible then talk to colleagues who have visited the site, examine brochures to obtain as clear a picture as possible and then add time for contingency.

Audit team selection

The Lead Auditor should select the audit team members, since the leader will be held accountable for performance of the audit: not the results but the effectiveness of the audit and the conduct of the auditors. The audit team should consist of:

- A Lead Auditor
- A number of auditors

[1] EN 45012 contains such tables.

- Specialists when necessary

- Interpreters when necessary

To carry out an effective and efficient audit the third party auditors should:

- Have knowledge of the industry sector

- Have knowledge of the particular products and services provided by the company

- Be nationally registered auditors

- Be able to devote the required preparation time

- Be available for the days required

- Be free of bias and influences that could affect their objectivity (see box)

The Lead Auditor should satisfy all the above criteria but does not need to have such an in-depth knowledge of the particular industry sector as other members of the team.

Specialists may be chosen for specific processes and so may not be in attendance for the complete audit duration. The specialist does not need to be a registered auditor and where this is the case a registered auditor should be in attendance.

Translators must be chosen by the Lead Auditor, not by the company; however, they must have a knowledge of the terminology employed in the particular industry.

> **AUDITOR INDEPENDENCE**
>
> - They have not held a position in the company being audited.
>
> - They have no financial interest in the company.
>
> - No members of their family or close friends are employed by the company.
>
> - They are not employed by a competitor.

In addition the Lead Auditor should choose people with whom he/she can work together. One must avoid having on the team:

- Conflict in the interpretation of ISO 9000 to the industry sector

- Conflict in the classification of nonconformities

- Incompetent auditors

- Clash of temperament

- Conflict in quality management ideology

The audit schedule

The audit schedule or programme defines which processes of the business are to be audited, when and by which auditors. The check lists will define what is to be examined in each of the processes. Some features are common to all audit schedules:

- Opening Meeting of 15 to 30 minutes on the first day

- Auditing periods for each process

- Lunch period of no more than 1 hour, better if it can be kept to ½ hr

- Auditor's review meeting once or twice each day

- Company feedback meeting where the audit duration is longer than 1 day

- Closing Meeting on the last day

The audit schedule is a guide to how the time will be divided and an indicator as to when managers of the selected processes ought to be available.

Inputs to audit schedule

The following inputs will be needed to prepare the audit schedule:

- Names of the selected auditors

- Floor plan

- Working hours of organization

- Availability of key managers

- The business process model

> **AUDITING STRATEGY**
>
> - Audit by process not by clause of standard.
>
> - For each process check compliance with the clauses of the standard that apply to that process.
>
> - A process may involve several departments and extend over a number of locations.

The processes to be selected should enable evidence to be obtained which will demonstrate achievement of the audit objective. If the audit is a full ISO 9001 audit then all elements of the standard should be addressed by these processes.

Preparing the schedule

A sample Audit Schedule is illustrated in Table 3-2. The clause numbers are not essential, being inserted purely for guidance. They will vary with the scope of the audit.

Date: 6 June 1994	Company: Top Class plc		
Time	**Lead Auditor** Cliff Hanger	**Auditor 2** Owen Yew	**Auditor 3** Bess Seller
09:00	Opening Meeting		
09:15	Business management • Purpose and mission • Customer expectations and needs (4.1.1) • Business plans/goals (4.1.1) • Quality policy, commitment (4.1.1) • Quality objectives (4.1.1) • Organization (4.1.2.1) • Resources (4.1.2.3) • Management review (4.1.3, 4.14)	Marketing • Publicity (4.5) • Tendering (4.3, 4.5, 4.16, 4.18) • Sales (4.3, 4.5, 4.16, 4.18) Order processing • Contract review (4.3, 4.5, 4.16, 4.18) Order planning • Quality/project planning (4.2.3, 4.5, 4.16, 4.18)	Production • Production planning (4.1.2.3, 4.9, 4.5, 4.16, 4.18) • Jigs tools and fixtures (4.5, 4.6, 4.7, 4.11, 4.16, 4.18) • Inspection and test planning (4.7, 4.10.1, 4.5)
12:00	Auditor's Review Meeting		
12:30	Lunch		
13:00	Quality systems management • Quality system representative (4.1.2.3) • Quality system development (4.2, 4.5) • Quality system maintenance (4.5, 4.14) • Quality system audit (4.17) • Quality records (4.16) • Quality system performance (4.1.2.1b)	Design • Design planning (4.4.2, 4.4.3, 4.1.2.2, 4.5, 4.16, 4.18) • Design organization (4.1.2.1, 4.4.3, 4.1.2.3, 4.18) • Design input requirements (4.4.4, 4.5) • Design activities (4.5, 4.18, 4.4.2) • Design review (4.4.6, 4.5, 4.14, 4.16) • Design output (4.4.5, 4.5)	Production • Machine shop (4.5, 4.9, 4.15, 4.16, 4.18) • Heat treatment (4.5, 4.9, 4.15, 4.14, 4.16, 4.18) • Coatings (4.5, 4.9, 4.15, 4.14, 4.16, 4.18) • Inspection (4.5, 4.10.3, 4.10.4, 4.10.5, 4.5, 4.12, 4.13, 4.14, 4.15, 4.16, 4.18, 4.20)
16:30	Auditor's Review Meeting		
17:00	Company Feedback Meeting		
Date: 7 June 1994			
09:00	Support services • Electrical equipment calibration (4.5, 4.7, 4.6, 4.11, 4.16, 4.18) • Mechanical equipment calibration (4.5, 4.6, 4.11, 4.16, 4.18) • Production Test laboratories (4.5, 4.10, 4.11, 4.16, 4.20)	Design continued • Design verification (4.4.7, 4.5, 4.16, 4.18) • Design validation (4.4.8, 4.5, 4.16, 4.18) • Design changes (4.4.9, 4.5) Support services • Test software and hardware (4.5, 4.6, 4.11, 4.16, 4.18) • Research & Development Test laboratories (4.5, 4.7, 4.10, 4.11, 4.16, 4.20)	Production continued • Assembly (4.5, 4.9, 4.15, 4.14, 4.16, 4.18) • Test (4.5, 4.10.3, 4.10.4, 4.10.5, 4.5, 4.12, 4.13, 4.14, 4.15, 4.16, 4.18) • Finished product stores (4.5, 4.15, 4.18) Support services • Maintenance planning (4.9, 4.5, 4.14) • Maintenance (4.5, 4.9) Order fulfilment • Picking and packing (4.10.3, 4.10.4, 4.12, 4.13, 4.15.2, 4.15.4, 4.15.5, 4.18, 4.20) • Dispatch/delivery (4.15.2, 4.15.6, 4.18) • Transportation (4.15.2, 4.15.6, 4.18)
12:30	Lunch		
13:00	Servicing • Service planning (4.9, 4.5, 4.16, 4.18) • Servicing (4.5, 4.9, 4.14, 4.15, 4.16, 4.18) • Servicing inspection (4.5, 4.10.3, 4.10.4, 4.10.5, 4.5, 4.11, 4.12, 4.13, 4.14, 4.15, 4.16, 4.18)	Procurement • Purchasing (4.5, 4.6, 4.16, 4.18) • Receipt inspection (4.5, 4.10.2, 4.10.5, 4.11, 4.13, 4.16, 4.18) • Raw material & component stores (4.5, 4.15, 4.18)	Installation • Installation planning (4.9, 4.5, 4.16, 4.18) • Installation (4.5, 4.9, 4.14, 4.15, 4.16, 4.18) • Installation inspection (4.5, 4.10.3, 4.10.4, 4.10.5, 4.5, 4.11, 4.12, 4.13, 4.14, 4.15, 4.16, 4.18)
16:00	Report Compilation and Feedback		
17:00	Closing Meeting		

Table 3-2 Sample audit schedule

The phrases used in the schedule need to indicate which processes will be audited and therefore will be to some extent dependent on:

- Where the work is carried out
- How work is divided between departments
- What the words are intended to convey

Contrary to what is stated in ISO 10011, an auditor should attempt to identify processes rather than departments as, depending on what particular departments do, the auditor may need to visit some departments several times during the audit as he/she follows audit trails through the process. Also, auditors should not be assigned specific elements as recommended in ISO 10011 as, invariably, several elements will apply to a process or audit area (see also under check lists).

The audit schedule should be flexible. It should not be treated as sacrosanct even when agreed with the company. If the audit schedule limits auditors to time-slots for particular departments, auditors will not be in a position to verify departmental interfaces so vital to verifying quality system effectiveness. A compromise is to identify processes and add department codes in order to indicate which managers need to be available. For example when indicating that Servicing will be audited at 1300 hrs on the second day, one could add '(S Eng, D Eng, QC, GR, S & R)', meaning Service Engineers, Design Engineers, Quality Control, Goods Receiving, Spares and Repairs.

Assigning auditors

Each member of the audit team is selected because of their particular expertise and so the Lead Auditor needs to assign auditors on the basis of their expertise in the processes to be covered. There is little to be gained from assigning an auditor with extensive software experience and no manufacturing experience to audit a manufacturing process and vice versa. It is normal practice for the Lead Auditor to cover business management and quality system management processes, as in these processes it is important for the Lead Auditor to establish the company's top management commitment to quality. Also, the company would expect that the Lead Auditor rather than another auditor interview the CEO. Another reason is that the Management Representative will probably know more about the quality system than anyone else and be very knowledgeable about the standard. It is important for the Lead Auditor to establish a good rapport with this person and gets his/her trust.

To verify that all clauses of the standard have been addressed, a matrix should also be prepared showing which clauses are to be addressed by which auditor in each process (see Table 3-3). 'X' indicates that the element is to be addressed and 'P' indicates which auditor is to take primary responsibility for collecting the evidence.

ISO 9001 Element	Lead Auditor	Auditor 2	Auditor 3
4.1	XP	X	
4.2	XP		
4.3		X	
4.4		X	
4.5	XP	X	X
4.6		XP	
4.7		XP	X
4.8		X	XP
4.9			XP
4.10	X	X	XP

ISO 9001 Element	Lead Auditor	Auditor 2	Auditor 3
4.11	XP	X	X
4.12	X	X	XP
4.13	X	X	XP
4.14	XP	X	X
4.15		X	XP
4.16	XP	X	X
4.17	XP		
4.18	XP	X	X
4.19			XP
4.20		X	XP

Table 3-3 ISO 9001 clause allocation matrix

Check list preparation

Purpose of check lists

Check lists serve to aid the memory and ensure that a systematic approach is taken to auditing. The bulk of information auditors use is stored in their long-term memory: the questions to ask, the processes to cover etc. However, as every organization is different, auditors need a means of helping them jog their long-term memory at the appropriate time when observing a process which they have not observed before. Where accuracy is paramount check lists ensure everything is addressed, even for those tasks which have been performed many times before. Here are some reasons for using check lists:

- Aids memory by indicating what has to be checked next
- Aids time management by indicating what has to be covered in a process
- Aids coverage of subject by causing a response to each requirement
- Aids the discovery of evidence by pointing the auditor in certain directions
- Aids collection of evidence by indicating what the auditor is looking for
- Aids investigation process by ordering the auditor's thought processes

Types of check list

There are four types of check list in common use, each with a distinct purpose. It is important to use check lists only for the purpose for which they have been designed, otherwise they will not reveal the information they were intended to.

Requirement check lists

Requirement check lists address each element, clause or individual requirement of the standard in the order presented in the standard.

One form of a requirement check list is a list of closed questions requiring a check against each to signify conformance or nonconformance. This type of list is suitable for use in conjunction with the auditee's documents for checking whether these documents adequately address the requirements of a particular standard or contract. They should not be used in conformance auditing, as they do not cause probing questions. However they can be useful after completing the audit to confirm that nothing has been overlooked. This type of check list has been found to be used by Certification Bodies and by companies as a supplier questionnaire. As it is structured around the elements of the standard it forces the auditor to audit element by element, when in practice this is not an effective method of auditing (see below).

Another form of requirement check list is to use open questions. These are used by quality system designers to record where in their system they have addressed each requirement. They can be used on conformance audits but tend to be too vague and ignore particular products/services/organizations etc. (A collection of such lists can be found in my *ISO 9000 Quality Systems Handbook*.)

Departmental check lists

There are two forms of departmental check lists: generic and specific check lists. Both forms are used primarily by internal auditors, as the information needed would not be readily available to external auditors.

Generic departmental check lists with closed questions can be used to help auditors cover everything but not in the sequence in which work may be carried out in a particular department. They are useful for generating specific departmental check lists.

Specific department check lists are useful for self audits but don't allow auditors to follow the process across department boundaries so are unsuitable for checking conformance with general standards.

A Tale of Misdirection

The company was known to Alan, the 'know-all' in the audit team, for he had been there some years ago. As all members of the team except Alan lived on the south side of Winterton, Alan proposed that he would meet them at the Little Chef at 0815hrs on the Winterton bypass and they would proceed together from there.

At 0830hrs, the team had still not assembled. Alan was seen driving in a northerly direction along the bypass by the other party who were driving south. At the next junction, Bill decided to turn north and follow Alan. Unfortunately Alan had also seen Bill and chose to do the same. When they eventually met up at the service station on the junction with Winterton Road, Alan was having an argument with the petrol pump attendant gesturing to a bare patch of land where there once stood a Little Chef, but apparently it had closed down several years previously.

It was now 0900hrs – the time set for the Opening Meeting. They were at least all together now but had to rely on Alan, whose memory was now questionable, to lead the way. Since the '80s, Winterton had expanded quite a lot and the industrial estate on the East side had been split into two, either side of the Bristol road. Needless to say Alan took the team to the wrong side but was fortunate enough to find a helpful hippie, who directed them successfully to the company.

They arrived flustered thirty five minutes late.

The moral of this story is:

If there is one thing as certain in life as death and taxes, it is change. Nothing stands still so be prepared for things to have changed!

Process check lists

A process is a sequence of tasks which combines the use of people, machines, methods, tools, environment, instrumentation and materials to convert given inputs into outputs of added value. A process will often extend across departmental boundaries and therefore a check list based on business processes will enable auditors to check the effectiveness of the controls employed by an organization.

A quality system is, in effect, a collection of processes which convert customer inputs into outputs that create satisfied customers. It follows therefore that auditing processes is a far better way to verify whether an organization has its operations under control than by auditing element by element, clause by clause of a standard that is only topic-based and not process-based.

The most effective way to verify compliance with the standard is to follow the processes from input to output, with diversions to investigate feedback loops, options and alternative routing of information or product. By following a process the auditor will follow a trail which invariably will lead across department boundaries.

A process check list can cover a single process, such as receiving inspection, or cover a group of related processes, such as procurement where one of the processes is receipt inspection (see later in this chapter and in Chapter 9 on check lists).

Process check lists are used for conformance audits to help auditors cover all aspects relative to a particular group of processes, regardless of departmental boundaries. Some can be generic but most have to be generated for each audit.

Process check lists that identify questions to be asked are useful for inexperienced auditors. Those that identify topics to be addressed are used by experienced auditors with topics acting as memory jogger. These allows auditors to follow processes regardless of who does what.

Flow chart check list

Flow charts can be either auditor generated or extracted from the organization's procedures. They can be used to verify conformance to procedure but are inadequate for testing conformance to standard or contract unless supplemented by requirement or process check lists.

Constructing the check list

In constructing check lists, they can be a series of separate lists, a consolidated list or divided into check lists for the documentation audit and other lists for the implementation audit. It is a matter of personal preference.

Documentation audit check lists

The most suitable type of check list for the documentation audit is a requirement check list. This would cover the following checks.

Requirement checks

The check list should contain each individual requirement taken from the applicable standards in full or condensed so that conformity/nonconformity can be indicated. Remember there are over 300 of these so this check list will be several pages.

Technology checks

A study of the organization's business will reveal the products, services and processes concerned. Using industry knowledge, a check list of the salient features, controls and risk areas can be generated. For example, an organization canning food should have a sterilization plant; an organization handling electronic components should have electrostatic handling facilities.

Implementation audit check lists

The most suitable type of check list for an implementation audit is the process check list, as it focuses on the way work is carried out in the business. An implementation audit check list would consist of the following:

Process checks	Tasks carried out in a process in the sequence that they are performed from input to output, in the form of an audit trail.
Product checks	Factors affecting the product or service being processed.
Measurement checks	Questions to test that the measurements required by the standard are carried out to verify product, service and process conformity and that the sensors are in place at the appropriate stage in the process.
Resource checks	Questions to test whether adequate resources have been provided to enable the organization's objectives to be achieved for a particular group of processes.
People checks	Questions to test whether the people managing, operating and verifying the process and its products are competent, capable, authorized etc.
Document checks	Questions to test whether the documents used and generated in the process meet the requirements of the standard and are under control.

Data checks	Questions to test whether the data used and generated in the process meet the requirements of the standard and are under control. (Documents have been separated from data as the type of controls are often different.)
Material checks	Questions to test whether the materials used or generated by the process are under control.
Tools/Equipment	Questions to test whether the tools and equipment used or generated by the process are under control.
Records	Questions to test whether the quality records used or generated by the process meet the requirements of the standard and are under control.
Policy checks	Items from the policies pertinent to the area or process that are not addressed by the governing standards.
Procedures checks	Items from the procedures that are implemented in the area or process.
Technology checks	Items from the technical standards, process specifications, industrial practices pertinent to the area or process.
What if checks	Questions to test the robustness of the system.

Compilation techniques

Many different techniques can be used for constructing a check list, either individually or in combination:

- Brainstorming, which sometimes throws up good ideas but tends to be a bit erratic and without logic

- Process analysis, which identifies inputs, outputs and what acts on the process to produce the results, correct errors and improve performance

- Requirement analysis, which ensures all key requirements are covered but overlooks the aspects peculiar to the business

- Flow charting, which can combine the advantages of brainstorming and requirement analysis but which needs actual procedures or process maps

- Evolution, which is what all check lists become but it is a poor way to start

If you haven't got the flow charts then the second best technique is to conduct a requirement analysis using a set of common factors. The procedure is as follows:

- Identify the process to be audited.
- Determine process boundaries to be confirmed.
- Address key factors[2] (see box).
- Locate the requirements of the standard which apply.
- Breakout the individual requirements of the clauses.
- Identify the topic to be covered.

KEY FACTORS
• Process
• Product/Service
• Measurement
• Resources
• People
• Documents
• Data
• Materials
• Tools/Equipment
• Records

By using the key factors to help identify the relevant topics to be covered you can create a very comprehensive check list (see example). The ten key factors could be reduced to Sayle's four factors[3] but subdivision of information into documents, data and records, and equipment into materials, resources and tools/equipment in my view provides a better prompt for the auditor. Also the addition of 'Measurement' is important to the purpose of the audit, namely that of establishing that operations are under control; and without measurement, there is no control.

In the example, only requirements from ISO 9001 have been selected but even so a list of 46 checks has been generated. As a stores does not provide an external service (i.e. a service to customers) this has been omitted. In practice, this list needs tailoring to delete those checks that are not relevant to the particular organization and to add the technology checks, procedure checks, policy checks and any other checks deemed appropriate. If the stores contains components, customer supplied product or finished product, then some of the checks will need to be made several times. If one were to audit against the standard clause by clause then when it came to audit the stores there would only be one clause on the check list, clause 4.15.3. Clearly there are more requirements which apply than those in clause 4.15.3.

---

[2] Allan Sayle in his book *Management Audits* identifies four key factors which he calls task elements, with further qualification as to correct type, condition, identification etc.

[3] The four task elements ibid. are Person, Item, Equipment and Information.

	Factor	Topic	Requirement Ref
		BUILDING A CHECK LIST FOR AUDITING STOCKROOMS	
		(Note the check list only includes checks derived from ISO 9001)	
1	Process	Designated storage areas	4.15.3
2		Authorizing receipt of product	4.15.3
3		Storage of product	4.15.3
4		Authorizing dispatch of product	4.15.3
5	Product	Identification	4.8
6		Inspection and test status	4.12
7		Handling	4.15.2
8		Packaging	4.15.4
9		Preservation	4.15.5
10		Segregation of conforming product	4.15.5
11		Segregation of nonconforming product	4.13.1
12		Identity of nonconforming product	4.13.1
13	People	Defined responsibilities and authority for stores personnel	4.1.2.1
14		Authority to initiate preventive action	4.1.2.1
15		Authority to identify problems	4.1.2.1
16		Authority to recommend solutions	4.1.2.1
17		Authority to verify solutions	4.1.2.1
18		Training of stores personnel	4.18
19		Qualifications of stores personnel	4.18
20		Understanding of quality policy	4.1.1
21	Documentation	Storage procedures	4.15.1
22		Procedure review and approval	4.5.2
23		Change control over procedures	4.5.3
24	Data	Deliverable product stock data	4.3.2c
25		Review and approval	4.5.2
26		Change control over stock data	4.5.3
27		Control procedure for stock data entry/change	4.5.1
28	Records	Training records for stores personnel	4.18
29	Materials	Control of packaging materials	4.15.4
30		Control of marking materials	4.15.4
31	Tools/equipment	Tools to monitor condition of stock	4.15.3
32		Tools to maintain storage conditions	4.15.3
33		Calibration of sensors	4.11.1
34	Resources	Adequacy of storage space	4.1.2.2
35		Assignment of trained personnel	4.1.2.2
36		Identification of training needs	4.18
37		Adequacy of staffing levels	4.1.2.2
38		Adequacy of handling equipment	4.1.2.2
39		Environmental conditions in storage areas	4.15.3
40		Control of environment	4.15.3
41		Identifying storage requirements	4.1.2.2
42		Identifying new storage controls, equipment, environment	4.2.3
43	Measurement	Stock condition checks	4.15.3
44		Intervals of stock condition checks	4.15.3
45		Audit of stores activities	4.17
46		Analysis of potential nonconformities	4.14.3

Table 3-4 Building a check list for auditing stockrooms

To use this type of check list the auditor needs to preface each topic with an appropriate phrase, such as 'What is the ...', 'How is ... carried out', 'Who is responsible for ...', etc. For further questioning techniques refer to Chapter 4.

Criteria for check lists

Whatever the format of the check lists, they should be fit for their purpose and therefore should:

- Identify what is to be examined
- Be concise
- Provide space for recording results
- Avoid closed questions unless the check list is to be used after the audit
- Not inhibit the line of questioning
- Cover the relevant requirements of the governing documents

Selection of guides

The selection of a good guide can make a lot of difference to the effectiveness of the audit. A guide who knows the company's processes and the people, and respects the auditor's right to gather information can eliminate unproductive interviews and searches for data. A guide who wanders off, takes the auditor the long way round, answers the questions instead of the auditee can be very frustrating. It is therefore important to acquaint the company with the role of the guide and give some guidance as to their selection.

The company should be requested to provide a guide for each auditor to:

- Take the auditor to the person or place to conduct the audit
- Resolve any problems encountered in the interview
- Assist in seeking information required by the auditor
- Witness and agree any findings reported by the auditor
- Assist the auditor in keeping to the schedule

Guides cannot be anyone who happens to be available. The company should be informed that in selecting the guides they should choose members of staff who:

- Know the personnel, the processes and their location so that time is not wasted

- Are not managers of the process audited so that staff are not under undue pressure

- Are not auditees at any time during the audit so that they can remain impartial

- Will take a keen interest in the audit so that they are alert to problems which require their assistance

- Do not have other commitments that day so that they are free to devote their time to their duties

It is not necessary for the guides to know the standard as they will not be asked to judge nonconformities by auditors. However, there was a period when the IRCA required the person signing the auditor's log to verify that the auditor had assessed all requirements of the standard. Clearly no one can sign such a statement without evidence. There are over 300 requirements in the standard so how will the Management Representative know whether or not all requirements have been checked? One means at his/her disposal is to employ the guides to maintain a discrete record of what has been verified and report back before the Closing Meeting. As the guides remain with the auditor throughout the audit, this is feasible but not practical. On the other hand it may not be in the interests of the organization to suggest that certain requirements have not been covered, so either the Management Representative exposes the organization to further checks and at the same time, embarrasses the Lead Auditor by disclosing that the audit has not been completed, or signs the log and says nothing. The former practice by the organization would certainly improve the quality of auditing with benefits to both sides, but it is likely that the latter approach would be taken. It was therefore not suprising that this condition has been removed by the IRCA.

Whilst the company selects the guides, the Lead Auditor does have the right to request they be changed in the event that their performance proves to be detrimental to the conduct of the audit.

Final confirmation with the company

It is desirable that the Lead Auditor confirms arrangements with the Management Representative prior to arranging travel details, just in case the situation has changed. Key personnel may be unavoidably unavailable, important customers may be visiting or there may be critical problems which, one way or another, make the timing inappropriate.

Details to confirm should include:

- The audit schedule

- The attendees at the Opening Meeting together with the agenda

- The availability of the managers of the processes to be audited
- Selection of the guides
- Audit team members
- Sizes of garments needed if protective attire is required
- Health and safety procedures
- Lunch arrangements
- Any other outstanding items from the Planning Check List and details on which there were concerns

Briefing of auditors

The Lead Auditor should give at least an initial brief and a final brief to the audit team.

The initial brief should be held once the auditors have been selected. The brief should include the following:

- Audit brief
- Purpose and scope of audit
- Pre-audit visit results
- Data and documents collected
- Draft audit schedule
- Allocation of auditors
- Success criteria
- Preliminary travel arrangements

The audit team should discuss these details and agree on the allocation of work and the success criteria, and undertake to study the quality system documentation, prepare the appropriate check lists and feed back any observations and requests for further data to the Lead Auditor.

Once the final details have been confirmed with the company, travel arrangements can be confirmed and the auditors give a final brief. At this final brief the audit preparations, including the check lists, should be reviewed and any outstanding issues resolved.

Auditor's materials

Auditors should carry with them a documentation pack and clip board or suitable folder with a hard surface on which to write. The documentation pack should include:

- The audit schedule

- A copy of the relevant assurance standard (ISO 9001, 9002 or 9003)

- A copy of the appropriate guidance standard (ISO 9000-2, ISO 9004-1, ISO 9004-2, ISO 9004-3 or other industry guide)

- The check list for the processes to be audited

- Report forms

- Note pad or book

- Pen and watch

- Protective clothing, where necessary, such as ear plugs, safety glasses, hard hats, safety shoes

- Signed confidentiality, health & safety statements where applicable

This done, the audit team is ready to commence the site audit.

Summary

This chapter has addressed audit planning and stressed its importance in the audit effectiveness. The following are the key issues covered:

◻ Before commencing audit planning an audit brief or audit requirement is necessary to authorize preparation to begin.

◻ A preparation check list is useful in ensuring all arrangements are made before commencing the audit.

◻ A pre-audit visit enables the auditor to make contact with the key personnel and gather sufficient information with which to plan the audit effectively.

◻ Establishing the correct purpose and scope of the audit is crucial to defining the boundary conditions for the audit.

◻ The success criteria has to be based on the factors that will determine system effectiveness and not arbitrary definitions of nonconformity.

◻ To require corrective action, failures have to be systematic rather than random.

◻ Examination of the quality system documents can provide useful data in planning the audit but unless all are provided, conformity with the standard cannot be thoroughly tested.

◻ Audits are more effective if carried out through the business processes rather than against the clauses of the standard.

◻ Audit duration is dependent on organization complexity, scope, competence and not only on size.

◻ The audit schedule has to be flexible to allow for changes.

◻ Auditing by process rather than department enables departmental interfaces to be checked.

◻ The Lead Auditor should always interview the top management to establish their commitment to quality and form impressions about the business.

◻ Check lists are no more than an aid to memory and not a list of all the questions to be asked.

◻ Check lists should be based on the areas or processes to be audited, subdivided into topics so that the audit follows a trail through the processes.

◻ Key factors can be applied to all processes to reveal sufficient information about the process, ensure consistency in audit techniques and conformity with the specified requirements.

◻ The selection of the guide is important to making productive use of auditing time.

◻ Final checks with the company before finalizing arrangements can prevent aborting the audit on site.

◻ Everyone on the audit team should be in no doubt as to what is required before commencing the audit and be equipped with the necessary materials to perform the task effectively.

Chapter 4

Conducting audits

The audit process

The audit process commences with the Documentation Audit and when any deficiencies have been resolved the audit team is then able to proceed with the Implementation Audit. This commences with an Opening Meeting on site and ends with an Auditors Review Meeting at which the findings are agreed. Depending on the number of audit days, this cycle will be repeated until auditing of all areas is complete. This process is illustrated in Figure 4-1.

Quality system documentation audit

Purpose

The purpose of the quality system documentation audit is to verify that the system will cause conforming product to be supplied and that adequate provisions are included to prevent nonconforming product being supplied. It should also establish that adequate provision has been made for meeting the applicable requirements of the standard.

Whether or not the documentation audit is carried out before the Opening Meeting, it remains a fundamental part of the whole audit[1].

---

[1] In QS 9000, the Quality System Documentation Audit is termed a Review and is part of the Assessment and can be carried out on or off site.

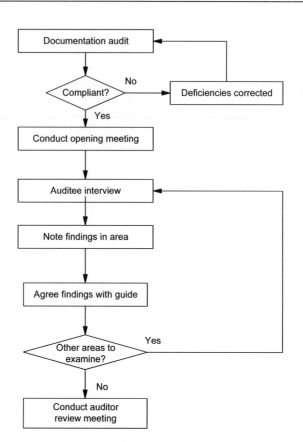

Figure 4-1 The auditing process

Location

The audit may be carried out on site where access to all the documents may be obtained or off site using documents provided by the company during the audit planning process.

There are pros and cons of on-site audits:

- The auditor can examine in depth all the relevant documents.
- The team can participate in the audit although it is time-consuming.
- Problems in understanding can be resolved on the spot.
- It incurs travel and subsistence expenses.
- If major nonconformities are revealed the visit may have been a waste of time.

There are also pros and cons of off-site audits:

- They are limited by the volume of documents which the company is willing to provide.

- Problems in understanding are more difficult to resolve over long distances.

- It is more difficult for the team to participate so the audit is often only performed by the Lead Auditor.

- When only the Lead Auditor carries out the examination, other auditors tend to ask questions for which they would know the answers had they studied the quality system documentation.

- They do not attract the travel and subsistence expenses.

- If a major nonconformance is found there is no waste of time as the on-site audit can be postponed to a time after the nonconformities have been corrected.

- Auditors tend to forget details and deficiencies in the documentation they examined several months previously.

During the pre-audit visit the composition of the documented quality system should have been established and so the decision on what documents to request for audit can be established.

The Lead Auditor should decide how the audit is to be carried out: on or off site, by him/herself or dispersed among the team. Getting the team involved is essential to some extent, as each auditor needs to know what provisions the organization has made so as to create the check lists and to avoid wasting time when on site.

> ### DOCUMENTATION AUDIT OPTIONS
>
> - Off-site as part of audit planning
>
> - Off-site as separate activity
>
> - On-site as part of pre-audit visit
>
> - On-site prior to implementation audit in each area
>
> - On-site prior to commencing implementation audit

If an on-site documentation audit is preferred, then the options are to conduct the on-site documentation audit during the pre-audit visit or during the formal audit. The advantage of doing the audit on the pre-audit visit is that it saves time during the formal audit but doesn't allow the other members of the team to examine the documentation. Whilst an on-site documentation audit during the formal audit will permit all team members to examine the documentation, it often causes delays. If the team members integrate the documentation audit with the implementation audit the chances are that the documentation will not be examined in sufficient detail to confirm compliance. The Lead Auditor should decide which approach is most appropriate as conditions will vary with each audit. The preferable approach is for the Lead Auditor to conduct the documentation audit on site as part of the pre-audit visit and for the team members to check the relevant documents against practice during the implementation audit.

If the documentation audit is to be conducted on site by the team, then prior to commencing the interviews in each area the auditor should request of the guide the applicable documents. The auditor should then spend some time confirming that they address the relevant requirements of the standard. Any findings can be written out on the spot and confirmed with the guide. Activities or evidence which need to be verified can be added to the check lists. When the documentation audit is conducted off site, there is always the danger that any detected nonconformities may be overlooked in the final report.

Carrying out the documentation audit

Many of the requirements of ISO 9001 will not be addressed by the Quality Manual if the manual only contains policies, so additional documents may be needed. However, the manual should indicate where such requirements are addressed, either through a matrix or in the text. Since the system should cause conformity and prevent nonconformity, the auditor should be seeking to establish that the documented provisions and no others will achieve this objective. In other words, is everything defined that is necessary to cause conformity etc. taking into account the education, training and competence provisions. If not, the system is incomplete.

The standard requires that procedures be prepared that are consistent with the requirements of the standard and the supplier's stated quality policy. This means that wherever the standard requires procedures, they should exist either as single documents, multiple documents or as part of a document. It also means that these procedures need to respond to the supplier's quality policy and objectives. If the quality objectives appear unclear, then clarification should be obtained from the company before proceeding, since it is not uncommon to find that objectives have been devised independently of the procedures and have not been taken into account when preparing the procedures.

To conduct the audit a complete set of quality system documents should be provided: documentation in Levels 1, 2, 3 & 4 or the policies, control procedures, operating procedures, standards and quality plans. In addition, a sample of current contracts covering the range of requirements the company is committed to meet should be provided.

The following checks should be made using the prepared check lists:

- That the documentation provided is under formal change control

- That the documents provided form part of the established system

- That all elements of the standard are addressed but not necessarily in the order of ISO 9001

- That there is a response to each relevant requirement of the standard either in the Quality Manual or in the documented procedures and standards (there are over 300 requirements so this is by no means a simple task)

- That adequate provision has been made to control the processes, products and services of the organization (see under check lists in Chapter 3)

- That the scope and applicability of individual procedures, standards, instructions etc. are compatible with the requirements of the standard

- That the provisions will cause conformity, i.e. cause the right things to be done

- That the provisions will prevent nonconformity, i.e. cause things to be done right

- That the policies and practices (procedures, instructions, standards, guides) are stated clearly and unambiguously

- That there is consistency within documents and between related documents

- That the policies and procedures or the quality plans address the requirements of particular contracts when contract requirements cause amendment, addition or removal of a company policy or procedure

If there are any discrepancies from the above checks then further enquiries should be made. If the Company cannot provide adequate documentation to cover the above then this is cause for a nonconformance report.

Typographical errors, spelling errors and other minor errors that will not cause incorrect action or decision should not be recorded as nonconformities. The standard does not require perfect documents. If there are many such errors or the documentation is untidy then bring it to the Management Representative's attention and do it verbally since to add these type of problems to the audit report will detract from the true purpose of the audit.

If an element of the standard has not been addressed and it is relevant to the operations of the organization then the Lead Auditor should discuss the matter with the Management Representative and agree whether the audit should continue or be postponed until the matter has been resolved.

Where there is some doubt that provisions are adequate, topics should be added to the relevant check lists so that clarification can be sought when conducting the implementation audit.

Remember that the audit should test conformity of all relevant requirements of the standard on a sample of the organization's operations. Checking that each clause is addressed is insufficient. Each relevant requirement of a clause has to be checked otherwise the audit is incomplete.

In examining the documentation, audit trails can be established from flow charts or procedural sequence and these should be added to the check lists.

Procedure adequacy

Establishing the adequacy of procedures is no easy task when sat at a desk. Their adequacy is more effectively assessed in practice. However, there are several styles that organizations may use. Some may be all text, others may have text supported by flow diagrams and some may have no separate text and be wholly in flow diagram form. There are also several approaches that can be taken. Some produce procedures that reflect how work is controlled, others produce procedures against each element of the standard and some produce procedures that define how work is carried out. There is no common approach, except than many produce procedures to correspond with the procedure requirement in the standard. Wherever the standard requires documented procedures to be established, they are produced. Whilst this approach may satisfy the standard it is my no means effective, as work processes will not match exactly the elements of the standard. Quite often what is classified as a procedure is in fact a policy statement or an instruction. The question that auditors have to wrestle with is 'When is a document a procedure or what constitutes a procedure'. Just because an organization calls a document a procedure does not mean that the document is a procedure. Procedures have to exhibit certain characteristics. These are stated in ISO 8402 and ISO 9004-1, but these alone are insufficient to separate an adequate procedure from an inadequate procedure[2].

---

[2] See *ISO 9000 Quality Systems Handbook*, Part 2 Chapter 2 for content of effective procedures.

Record adequacy

Record formats should be examined when auditing the documented quality system as they represent the provisions the organization has made to gather objective evidence that operations are under control. Access to record formats is needed to verify implementation of certain requirements of the standard. However, there are records and quality records. Not all records are in fact quality records.

Sometimes the standard specifies what must be included in the quality record, such as the acceptance criteria in the requirement for inspection and test records of clause 4.10.5. However, in most instances where clause 4.16 is referenced, there are no requirements for the content of records so the auditor has to judge whether the record being examined is adequate. As with procedures, just because an organization calls a document a quality record does not mean that it is a quality record. Some organizations list all their records as quality records, even the administrative records such as time sheets, travel logs etc. One has to accept that few organizations will apply the standard as it was originally intended. The advice they are given may well conflict and may result is some weird solutions. The auditor needs to establish whether the documents offered are in fact quality records and for them to be adequate they have to exhibit certain characteristics[3].

Documentation audit report

The report of the documentation audit can either be a written report listing all the discrepancies found in the documentation examined or the Audit Findings sheet (Figure 5-7 in Chapter 5) can be used. Where the audit takes place off-site or during the pre-audit visit on-site, then a written report is the more usual method. When writing the report, the findings should be presented against each clause of the standard; however, care should be taken to avoid making general statements out of single checks as the system may not be compliant in all areas where the requirement applies. For instance, it is inappropriate to claim the system complies with the document control requirements if the only documents examined are the quality system procedures. There are many other documents that need to be under documentation control so the scope and applicability of procedures needs to be carefully examined.

A summary of the results of the documentation audit should be recorded in a Compliance Matrix (see Figure 5-6), which should merely record against the clause number of the standard, whether the provisions are adequate to demonstrate confor-

---

[3] See ISO 8402 for a detailed definition of 'quality records'.

mance with the standard. This matrix should be included in the final audit report as the implementation column will be completed later on in the audit. Where requirements have definitely not been addressed in the quality system documentation then nonconformity statements should be prepared quoting the relevant clause of the standard (see Chapter 5). These can be withdrawn later if evidence of adequate documentation is found during the implementation audit.

Remedial action

A period of up to three months is usually allowed for the company to respond to the findings of the documentation audit. However, at the time of the audit, the auditor will not know whether the omissions in the documentation were merely omissions or indicate an absence of a provision in practice. For the former case, documenting existing practices which meet the standard can be accomplished in less time than introducing new practices. It is therefore important for the Lead Auditor to establish whether the omission is one of documentation only or whether there are no provisions at all in place. In the latter case the implementation audit should be delayed until such time has elapsed for evidence of implementing the new practices to be accumulated.

On receipt of the documentation audit report, the company is expected to take remedial action to correct the discrepancies and either resubmit the revised documents or indicate what action has been taken. This information should be examined and if found acceptable, the implementation audit can proceed. Where minor discrepancies remain, it is often prudent to resist requesting further changes as the implementation audit may reveal that the problem requires more than a change to the documentation.

The Opening Meeting

Purpose

On arrival at the site the auditors should assemble for the Opening Meeting. The purpose of the Opening Meeting is to establish communication with the company, confirm the arrangements and resolve any difficulties before the audit commences. With internal audits an Opening Meeting is usually unnecessary since the auditors are known by the staff; however, some introductory meeting is preferable if only to establish that nothing has changed since agreement of the date and scope of the audit.

Attendance

The Opening Meeting should be attended by all the auditors together with representatives of the company. The representatives of the company should include as a minimum:

- The Chief Executive or Managing Director. It is important that the CEO attends in order to demonstrate commitment. The Lead Auditor will be interviewing the CEO immediately following the meeting. In this way, presence ensures availability. If the CEO is unavailable a deputy should be present.

- The Management Representative for the Quality System. It is important that the Management Representative agrees to the scope and the plan for the audit and which members of the company are authorized to agree the findings.

- Managers of each of the areas to be audited. It is important that the managers agree the scope of the audit and are aware of which auditors will be conducting the audit in their area at which times and in which sequence. It gives them the opportunity to alter the sequence if conditions are inconvenient.

- The guides. It is important they understand their role in the audit and meet the auditors they will be guiding.

Opening Meeting agenda

The agenda for the Opening Meeting is given in the box opposite. The duration of the meeting should be no more than 15 minutes so each item should be dealt with quickly.

The sequence of items is fairly important in order to convey a positive impression. One wouldn't start by defining nonconformities for instance and it is prudent to give the company the opportunity to introduce their attendees before launching into the audit process. Most of the items should be presented in order to confirm details agreed during previous contact with the company.

To set the scene, the purpose and scope of the audit should be uppermost in the

> ### *AGENDA FOR OPENING MEETING*
> - Introduction of the audit team
> - Introduction of company personnel
> - Purpose and scope of the audit
> - Audit schedule
> - Auditing process and reporting procedures
> - Success criteria
> - Appointment of guides
> - Agreement of findings and nonconformities
> - Heath & safety
> - Facilities needed
> - Confidentiality
> - Disclaimer
> - Questions

agenda and confidentiality held until near the end. Finally, questions can be invited from the company. Each of these items is discussed in more detail below.

Conducting the Opening Meeting

The Opening Meeting can either be a non-event or a traumatic experience. The Lead Auditor should have met the key staff beforehand but audit team members will be unknown to the company. It can be a time for the company to size up the audit team and form an opinion as to whether they are real professionals or nit pickers. Some staff may attempt to test the knowledge of the auditors selected for their areas by asking whether they are qualified, although it may seem a foolish thing to do at such a late stage. Others may seek information to be sure that there is no conflict of interest. Where the company is seeking third party certification, they may be keen to show the audit team they have worked very hard developing the quality system and believe it to be bringing positive benefit. Others will declare that they have no customer complaints and ISO 9000 is only formalizing what they have been doing for years.

In preparing for the Opening Meeting the Lead Auditor should establish what is to be said by whom against each item on the agenda. It is important to expose each member of the team to the company so that they can at least form some opinion of each team member. The silent auditor can be very off putting. Either he/she could be weak or devious in their eyes. However as soon as an auditor speaks he/she may allay or strengthen their concerns.

The introductions

This presents no problem if there are few people at the meeting. The Lead Auditor can introduce him/herself and leave his team to introduce themselves to the company staff present as they walk into the meeting room. Where this can get out of hand is where there is a large audit team and large representation by the company. If this situation arises, the Lead Auditor should take the lead and introduce him/herself and the members of the audit team and then invite the company to introduce themselves. The practice of everyone shaking hands with everyone else at once can be chaotic. When the auditors introduce themselves it is a good idea if they briefly explain which areas they will be auditing but this may be done later when explaining the audit schedule.

When numbers or names warrant it, the Lead Auditor should pass around an attendance sheet so that names are recorded accurately. The names of key personnel will be needed for the audit report so having them recorded at the Opening Meeting saves any embarrassment later due to incorrect spelling of names.

When the company introductions are over, the CEO may wish to say some words about the company. This could be brief but may be up to an hour duration with overhead slides and video presentation. The Lead Auditor should politely ask how long it will take as only 15 minutes have been allowed in the plan for the Opening

Meeting. If the CEO says it will be only five minutes then the chances are it will last only three minutes. If it goes beyond five minutes then some appropriate body language from the Lead Auditor may do the trick and draw the presentation to a close. It is important to remember that the audit team are guests and whether second or third party, they are present only with the agreement of the company. Therefore always be polite. The Lead Auditor could suggest that the presentation be given after completing the agenda of the Opening Meeting or that he / she has already viewed the presentation during the pre-audit visit and taken account of the valuable information when preparing for the audit.

Confirming the purpose and scope

More problems occur over the purpose and scope of the audit than anything else. Although this will have been checked and may have been altered with the first contact with the company and checked again during the pre-audit visit, everyone seems to have a different understanding of what is being audited against what standard and for what products and services. Those at the Opening Meeting may not have been involved in the discussions establishing the scope and so may be unaware of the limitations. The Lead Auditor should read out the purpose and scope statement in order to avoid any slip up. The same statement should be read out at the Closing Meeting so as to avoid any unintentional misunderstandings. Examples of purpose and scope statements are given in Chapter 3.

Confirming the audit schedule and availability of key personnel

Anything may have occurred since agreeing the schedule with the Management Representative. It is therefore essential to confirm the areas to be audited in the sequence they will be visited. The Lead Auditor should present the schedule or request a member of the team to present it. Copies of the schedule should be handed out or, if facilities are available, an overhead transparency could be prepared. The times for lunch break, auditor meetings, feedback meetings and the Closing Meeting should also be confirmed. The Lead Auditor should be prepared to alter the plan if necessary to accommodate minor changes. If the changes are significant then it may be necessary to re-schedule the audit for another date when the managers are available and the audit will cause less disruption. If this should happen then it indicates from the outset that the Lead Auditor is either incompetent or that there is a serious communication problem which may exist not only with the plan but with many other aspects of the company's quality system.

The Lead Auditor should confirm who will attend the company feedback meeting and Closing Meeting and escort the team to lunch if taken on site.

Explaining the audit process and reporting procedures

If the audit process is new to the company then an explanation to those present at the Opening Meeting is essential to avoid any misunderstandings. Whilst the audit should test the system in actual operation, as this cannot be done without interviewing staff, it will inevitably cause some disruption. It is useful also to establish what briefing and preparation the company has given to the staff about the audit so that the team may be prepared for the extent to which they have to explain the purpose of their visit to each person they interview.

The Lead Auditor should explain the following:

- The auditors will be seeking conformance to the requirements of the standard by interviewing members of staff responsible for work which affects product or service quality.

- They will examine samples of work and associated documentation to establish that the documented system and its implementation conforms with the requirements of the standard and any findings will be recorded.

- The auditors will agree the findings with the guides and these will be classified as either observations or nonconformities.

- When nonconformity statements have been prepared, normally at the end of each day, they will be presented for agreement to the Management Representative.

- Information will be passed to management at daily feedback meetings and all the results of the audit reported at a Closing Meeting on the last day when a hand-written report will be provided and a date agreed for the submission of any corrective action proposals.

- The company will be informed of the recommendations which the team will make to the Certification Body before leaving the site.

Explanation of success criteria and definitions of nonconformities

After explaining the audit process, the Lead Auditor should explain the success criteria and the definitions that will be used to determine any nonconformities. It is wise to explain these things beforehand to avoid any surprises and overcome any misunderstandings. Also treating this as success criteria rather than nonconformity criteria creates a more positive impression. The success criteria are those factors which indicate that the quality system is effective. In making this presentation the Lead

Auditor should explain that the team will be seeking conformity and in so doing may come across nonconformity. It is not the purpose of the team to find nonconformity although invariably some nonconformities will be found. If the Leader focuses on nonconformity it conveys the impression that the audit is a negative rather than a positive activity. The Leader can then read out the definitions of a nonconformity, a major and minor nonconformity, observations and the possible outcomes (see Chapter 3).

Confirmation and role of the guides

The Lead Auditor or a team member should request the company to indicate who are to act as the guides for the duration of the audit and confirm that they are suitably qualified. The role of the guides should be explained so that as they are present they may seek clarification if in any doubt as to their duties. See Chapter 3 for details of the qualifications and role of guides.

Agreeing findings and nonconformities

The Lead Auditor should confirm that the guides are authorized to agree findings and request the company indicate who is authorized to agree nonconformities. It is not necessary for the guides to know the standard but the person agreeing to nonconformities should have a sound knowledge of the standard and will usually be the Management Representative.

Confirmation of health and safety issues

If the company's health and safety notices have been read beforehand then it should be confirmed that they have been understood and that the team will adhere to the company's health and safety procedures. In some companies, signed statements are necessary for insurance purposes and these should be handed over before closing the meeting. If any notices were unavailable at the planing stage then the Lead Auditor should enquire as to the nature of any health and safety matters that will affect the team and provide the necessary assurances. Where protective attire is necessary its availability in the sizes required should be confirmed.

Facilities required

The Lead Auditor should confirm that areas are available for the team review meetings, the company feedback meetings and Closing Meeting. The team review meeting room needs to be centrally located since access to the Management Representative will be required. However, it need not be as large as the room reserved for the Opening and Closing Meetings.

Any transportation needed for auditors to reach the audit areas should be agreed.

Confidentiality statement

The Lead Auditor should declare that any information obtained during the audit will remain confidential and not be disclosed to a third party without the prior agreement of the company. In some cases the company may request the team to sign a Non-Disclosure Agreement. Submission of previously prepared agreements signed by the audit team will indicate sincerity.

Disclaimer

The Lead Auditor should explain that the audit is based on a limited sample of operations and although conformance with all relevant requirements of the standard will be tested, other nonconformities to those reported may exist.

At the end of the Opening Meeting the auditors should locate their guide and commence the audit without delay.

Implementation audit techniques

The primary aim of the auditor is to discover facts and these can be obtained by a combination of observation of operations in practice, by interviewing personnel responsible for carrying out such operations and by examining documentary evidence such as procedures, plans, specifications and records. The auditor obtains much of the information through interviewing people and this is not a simple task, especially when such people would rather not disclose their weaknesses or that they had been told not to do so by their manager.

Auditors should be looking for conformity not nonconformity. If their objective were to only find nonconformity, their task may be over soon after the audit commences. The audit objective is unlikely to be that of determining the extent of nonconformity. It is more likely to be that of verifying the extent of conformity. In seeking conformity they may stumble across nonconformity. There will always be some nonconformities or random errors. What the auditor should seek to establish is whether the noncon-formity is an isolated error or is a symptom of an ineffective quality system. Isolated errors are often caused by human fallibility and may be inadvertent or caused by a lack of training. The auditor should establish which is the cause and only report those where a lack of training is evident (see later under *Reporting*). This requires that the auditor must at all times keep the audit objective in mind. Auditors should work separately, not in pairs, unless one team member is a trainee or a specialist without audit skills.

Establishing conformity

The audit of the quality system documentation established that the company had or had not made adequate provision for causing conformity and preventing nonconformity and meeting the requirements of the standard or contract. The implementation audit serves to confirm that these provisions have been implemented by the company wherever they apply.

The company's documented policies and procedures may go beyond the requirements of the standard or contract and may impose specific policies and methods to satisfy a particular requirement of the standard or contract. Some auditors adopt the principle that the company should document what it does and do what it documents. This approach may or may not produce objective evidence of compliance with the requirements of ISO 9000 or a particular contract. ISO 9001 does not dictate how a particular requirement should be met although there are some cases where a series of activities is prescribed for meeting a requirement (clause 4.13 is one example). The auditors task is to establish through the documentation audit the method prescribed by the company for meeting a requirement and then verify that this method is followed. The auditor should not seek to verify methods which do not respond to a requirement of the standard or a contract. There are three situations for which judgement is needed:

- Where the documented system includes areas not addressed by the standard e.g. marketing, occupational health and safety, unintended product.

- Where the documented system prescribes a particular solution to a requirement of the standard and does not make provision for any alternative, e.g. use of an Approved Suppliers List for selecting suppliers, use of Taguchi techniques to determine design characteristics.

- Where the documented system defines conditions that the standard does not require, e.g. the intervals of internal audits, dates on documents in addition to revision level.

In the first of these situations the topics are not addressed by ISO 9001 so should not be included in an ISO 9001 implementation audit unless they are the business of the company. Obviously marketing would be addressed when auditing a company that offers marketing services and safety would be addressed if the product happens to be a live person as in the case of a hospital. If they are addressed by other standards such as QS 9000 or specific contracts then they should be included in the audit.

In the second case, not following a prescribed method when no alternative has been specified is indicative that the activities are not being controlled even though these methods are not required by ISO 9001. If alternatives are specified and appropriate controls are in place then there is no nonconformity.

In the third case, judgement is more difficult as the reason for the provision needs to be established. The auditor needs to establish why audits are scheduled monthly since it is not a requirement of the standard. It could be an arbitrary stipulation or a constraint in order to ensure an objective is met, such as the audit of the complete system once each year. The auditor may reveal that audits are being performed when resources are available rather than monthly. Is it a nonconformity? The answer is not simple. The standard requires audits to be scheduled on the basis of the status and importance of the activity. Unlike the requirement for management review in clause 4.1.3, the audit requirement does not require the frequency of audits to be defined. If the audits carried out demonstrate that they were scheduled on the basis of status and importance then the nonconformity does not concern the audits but the audit procedure since it does not cause audits to be scheduled on the basis of status and importance. The auditor has to use judgement to establish whether the actual frequency is effective. It could be that the plan was too ambitious and that any requested corrective action would be to change the plan rather than follow it.

Auditors have to be careful not to cause companies to conform to their procedures when their procedures are impractical. If a nonconformity statement states that practice was not in accordance with procedure then the procedure has to be correct so that the corrective action changes the practice. If the practice is correct and the error lies with the procedure then the nonconformity statement should cause corrective action to change the procedure.

The interview

Who to select for interview

Selecting the right person to interview can save a lot of time. The auditor must know what he/she is trying to establish before interviewing anyone. You need to talk to the manager responsible for a particular process to establish that they have their operations under control, You also want to interview members of staff in the area to verify that they not only know what they are supposed to do but are complying with the documented policies and practice to the extent that they satisfy the requirements of the standard. So you ask the guide to take you to the person responsible for a particular process or a particular activity or decision, if the responsibility for the process is shared between managers. Some processes are contained in a particular area such as design verification, goods receiving or finished product stores although the managers may be located elsewhere. With an operation that is carried out by a number of people then you would ask the manager if they all performed the same task. If they did, then ask the manager if you could interview a few of them, the number being proportional to the number of people so that you obtain a balanced result (unless of course there are several hundred people all doing the same thing). If the people perform different tasks, then you would need to establish which tasks were more important than others to product quality and interview each of these.

You also need to establish how long people have been doing a particular job and include in your sample those who are relatively new, to verify the extent of training given, and those who have been doing the job for many years to verify that they have adapted to the quality system. Old habits die hard and the old timers may be reluctant to change. This attitude may be passed on to the new recruits so working relationships are a good indicator of whether the system is being effective. On the other hand, the old timers may have been trying to change the system for years and welcome the changes, whereas those who have only been in the organization for two years and may not stay much longer, may be more concerned with their career than with policies and procedures.

You can be guided by the guide as to the best person to talk to, i.e. the most knowledgeable, and by all means interview that person, but also seek out others to test whether the system is used by everyone. Don't be diverted away from the new hire as you need to test the induction procedures. Also don't be led by the guide into areas you haven't planned to examine or to people you haven't planned to interview. The guide is more of an escort than a tour courier.

Interview format

Before starting an interview the auditor should:

- Have the appropriate documentation pack available

- Know what is to be established i.e. the subject of the inquiry

- Have read the policies and procedures which apply to the area sufficient to be aware of the processes which apply and have added items to the check lists of things to check as a result

During the interview the auditor should:

- Ask open questions to reveal facts

- Ask probing questions to investigate the facts

- Avoid emotive questions, trick questions, leading questions and multiple questions

- Listen intently

- Exhibit correct body language

- Be observant and inquisitive

- Take notes

- Check evidence with the standard and/or auditee's documentation

- Establish the root cause of deficiencies

- Request sight of the objective evidence

- Give positive feedback

- Avoid giving advice

- Remain calm, cool and objective

- Ask closed questions to confirm the facts

- Thank the auditee for his/her help and co-operation

> **INTERVIEW FORMAT**
>
> - Introduce yourself.
>
> - Explain the reason for the interview.
>
> - Ask open questions to reveal information about the subject of the inquiry.
>
> - Respond to answers with probing questions.
>
> - Seek objective evidence of the facts as reported.
>
> - Check results against the standard and/or policies and procedures.
>
> - Ask closed questions to confirm the facts.
>
> - Record findings.
>
> - Thank auditee for their help and co-operation.

Each of these points is covered in more detail later in this chapter.

After the interview the auditor should:

- Confirm findings with the guide

- Make alterations to his/her notes

- Move on to the next area

Interview objective

When conducting an interview the auditor should keep in mind certain objectives:

- That the organization has got its operations under control

- That the controls conform to the relevant standard

- That there is objective evidence of conformity and, where applicable, nonconformity

- That operations remain under control when there are problems

In establishing that operations are under control, the auditor should verify:

- That the auditee knows of the requirements his work has to satisfy

- That these requirements are those defined by the company's quality system

- That the auditee has access to the documentation necessary to perform the work

- That the plans and procedures are adequate to cause conformity and prevent nonconformity

- That the specified requirements are followed

- That the auditee's work is subject to verification prior to release

- That the auditee has the ability to change the process should the requirements not be met

- That provision has been made to prevent the unintended use of noncompliant work

Fact finding

To reveal facts about a process the auditor should:

- Test for conformity on one sample.

- If conforming, take 2-3 more samples if more samples exist.

- If conforming, establish if samples are representative of subject. Samples taken from the same department/contract/product/project may not be representative of other departments/contracts/products/projects.

- If not representative, take other samples in other locations, processes, projects, contracts or products.

- If one sample is nonconforming, establish whether all samples would be nonconforming by virtue of the established practice or procedure.

- If all samples would not be similarly affected, take 2-3 more samples to establish that it is not an isolated case.

- If these are nonconforming establish the significance with respect to its affect on quality.

- Obtain suggestions from auditee on possible root causes of nonconformity. Possible causes are inadequate procedures, plans, specifications to ensure consistency or resources and management commitment.

- Record details of document and product identity, location and any other details to enable others to locate the incident. Both auditee and future auditors may need to locate the problem some time later. Don't leave them with a conundrum.

Seeking objective evidence

Objective evidence is defined in ISO 8402 as 'information which can be proved true, based on facts obtained through observation, measurement, test or other means'. Much of the objective evidence gathered during an audit will be from documentation either in the form of policies, procedures, specifications etc. or as records. However, evidence does not have to be documented for it to be objective. An auditor can prove information to be true by observation, i.e. watching someone perform an activity. This is where the guide is most useful as he/she can testify as to the truth of what the auditor observed without there being documentary evidence.

ISO 9001 requires quality records to be maintained for certain activities but not all activities. Most of the activities for which quality records are required are critical to demonstrating that products conform to specified requirements. There is no requirement in ISO 9001 for the supplier to prove through documentary evidence that the requirements have been met. Processes may involve many activities, not all of which need to be recorded. If they were, there would be a shortage of trees or electronic storage capacity! The auditor can verify that processes are under control by sampling at discrete points in the process and sampling the output. One does not need documentary evidence that every stage in a procedure has been followed to demonstrate that it has or has not achieved its purpose. The objective is to obtain sufficient confidence that the requirements of the standard are being met rather than to seek documentary evidence of compliance with 100% of the requirements.

Following a trail

It is better to follow a logical trail towards an objective than to shoot off in all directions aiming wide of the objective, asking questions at random without any apparent purpose. It puts the auditee off balance and does not make effective use of time. The examination of quality system documentation should have revealed the trails to be followed. These trails should focus on the key processes but may often deviate to investigate the exceptions, unusual situations and support functions.

In following a trail, ask questions to establish where the information or the product comes from to start the process. Then ask what they do next and so on until the final stage. Then ask where the outputs go to and with what documentation. Along the route, ask What if? type questions to uncover whether provisions have been made to deal with changes in the flow and prevent uncontrolled deviation. Some areas to explore are:

- The different pathways for routing product and information

- What happens if a revision to a document is received after work has commenced, thereby testing the change control provisions

- What happens if the instructions are incomplete, unauthorized or the input is not compliant with the input requirement, thereby testing discipline and commitment

- What a person would do if instructed by someone else who is not his/her supervisor to release product or information that was not complete, thereby again testing discipline and commitment

What to ask the manager

Every manager takes responsibility for a particular contribution to the business and has power to influence the way operations are carried out and hence affect quality much more than an individual operator. Managers will be the type of people who will lead their staff towards the organization's purpose and mission or away from it. The auditor's task is to establish which type of manager they are interviewing. In addition there are certain requirements of the standard which need to be verified with any manager regardless of the function they manage:

- What involvement have they had in the development of the quality system (clauses 4.1.2.1 and 4.5.2) – a question that tests understanding

- What benefits have been gained as a result of using a quality system (clause 4.2.1) – a question that tests commitment

- What is the extent of their accountabilities (clause 4.1.2.1)

- What are their responsibilities and authority (clause 4.1.2.1)

- In what documents are their responsibilities defined and documented (clause 4.1.2.1)

- Who do they interface with outside their function to perform their job (clause 4.1.2.1)

- Where are these interfaces defined and documented (clause 4.1.2.1)

- What requirements/targets are they required to meet (clause 4.1.2.1)

- How do they plan the work load needed to satisfy these requirements (clause 4.1.2.2)

- How do they establish the resources they need to discharge their responsibilities (clause 4.1.2.2)

- How to they assign work to their staff (clause 4.1.2.2)

- How do they establish that their staff are carrying out their work as directed (clause 4.1.2.1)

- What feedback do they receive on their performance and that of their staff (clause 4.1.2.1)

- What measures are taken to detect opportunities for improvement in performance (clause 4.1.2.1)

- How do they identify training needs (clause 4.18)

- How did they learn to do the job (clause 4.18)

What to ask the staff

The questions to put to the staff are somewhat similar but more oriented towards their operations. Apart from specific questions relating to particular processes, there are certain requirements of the standard which need to be verified with any employee regardless of the job they perform:

- What are their responsibilities and authority (clause 4.1.2.1)

- In what documents are these responsibilities and authority defined (clause 4.1.2.1)

- In what way does the quality policy affect what they do (clause 4.1.1)

- What would they do if they felt they were not competent to carry out the work required (clause 4.1.2.2)

- What would they do if the instructions they received could not be implemented (clause 4.1.2.1c)

- If problems arose in their work, what would they do about them (clauses 4.1.2.1b and 4.1.1)

- What feedback do they receive about their performance (clause 4.1.2.1a)

- What would they do if they felt that the processes in which they were involved could be improved (clauses 4.1.2.1c and 4.1.1)

- How did they learn to perform their job (clause 4.18)

- What training and education is provided for them to learn to do their job more effectively (clause 4.18)

- What involvement did they have in the development of the quality system and its procedures (clauses 4.2.2 and 4.5.2)

- How has the quality system benefited their job (clause 4.2.1)

What not to ask

Certain questions may appear impertinent or insensitive, such as:

- 'What makes you qualified to do this job?' Instead ask: 'What qualifications are required for someone to do this job?'

- 'What training have you had to do this job?' Instead ask: 'How did you learn to do this job?'

- 'What is the quality policy?' Instead ask: 'How does the quality policy affect what you do?'

- 'Who told you to do that?' Instead ask: 'What instructions were you given to carry out this work?'

- 'May I see your procedure for handling customer supplied product?' Instead ask: 'Does your customer supply you with any product?' Then if the answer is 'No' procedures are obviously not needed.

- 'Why is this equipment not calibrated?' Instead ask: 'What is this equipment used for?' Then if it is used to determine the acceptability of product, its accuracy should be verified. Also this question presumes that the equipment is not calibrated. Equipment that does not carry a calibration label may in fact have been calibrated. The records need to be examined to verify this so ask: 'Where are the records which indicate that this equipment has been calibrated?'

- 'Why has this out of date document not been removed?' Instead ask: 'What is this document being used for?' The fact that the auditor could tell it was out of date indicates that it carries an identity.

Being observant

Auditors need to be observant. More like a Sherlock Holmes than a Walter Mitty. When being shown around areas or processes or when being shown documents and products, aspects to look out for may include:

Documentation

- The status of the documents being used (current or obsolete, clean or dirty, complete or incomplete, authorized or unauthorized changes)

- The validity of diagrams, forms or other extracts from documents posted on office walls or machines (current or obsolete, clean or dirty etc.)

> **OBSERVE**
> - Documentation
> - Areas
> - People
> - Products
> - Equipment/tools
> - Data
> - Materials

- The availability of documentation to personnel (proximity of location to point of use, accessibility by personnel during hours of work, whether locked away etc.)

- The awareness of personnel of the documentation they are required to use

- Evidence that the documentation is in use and not collecting dust on the shelf

- Evidence that the documentation adequately reflects the operations being described

Areas

- The general level of housekeeping. Poor housekeeping may be a hazard to product or cause loss or deterioration to documents. An untidy office is not necessarily indicative of ineffective control, just as a tidy office is not necessarily indicative of effective control. Check the competence of the people in the area first.

- The layout and its relationship to the flow of product and information. Displaced activities can lead to people taking short cuts to avoid delays, to avoid walking the distance etc. The more movement the more handling may be involved, thereby creating potential hazards.

- Bottlenecks where product or information is held awaiting processing. These may indicate inadequate resource provisions, low process capabilities etc. Bottlenecks may put pressure on those doing the work or releasing product or information to act in inappropriate ways.

- Notices, when were they posted, are they relevant, do people obey them etc. A notice on zero defects may be from an abandoned campaign rather than current policy. Ear defenders may not be being worn and when operators can't hear instructions, this can affect product quality.

Product

- Condition of product (clean or dirty, undamaged or damaged, leaking etc.) Unclean product is not necessarily indicative of poor control, it depends on the nature of the process. Rusty castings are not nonconforming, it is part of the ageing process, the rust is removed in the shotblasting process.

- Identity of product (part markings, modification status, inspection status, serial numbers etc.) Product carrying no identity markings may be identified by their location or their shape so don't jump to conclusions.

- Handling and segregation of product, containers, protective measures, lifting provisions etc. How it is stored where it is produced, how it is handled by the operators may indicate a lack of care and hence no commitment to quality.

- Packaging of product (component parts, in-process and final product). Know industry practices. Electrostatic protection applies to certain electronic components, not all.

- Warning notices, when relevant, of hazards to product and personnel. Are fragile products protected, are containers the right way up?

Equipment/tools

- Use. Confirm what it is used for before checking other aspects as it may not affect quality of deliverable product or service.

- Condition of equipment/tools (clean or dirty, undamaged or damaged etc.) Unclean equipment may indicate frequent use or poor housekeeping. Waste material may obscure operating instructions, gauges etc.

- Identity (type numbers, serial numbers, version etc.)

- Status (calibration, verification, modification status etc.) Not all measuring instruments will need calibration, it depends what they are being used for.

- Operating instructions where necessary either on equipment or close by. Are they registered, current and under control?

- Warning notices when relevant of hazards to product and personnel

Data

- Use. Confirm what it is used for before checking other aspects as it may not affect quality of deliverable product or service.

- Validity (approval, status).

- Integrity. Restricted access for changing data.

- Recency. When was the accuracy last confirmed.

Materials

- Use. Confirm what they are used for before checking other aspects as they may not affect quality of deliverable product or service.

- Condition (clean or dirty, undamaged or damaged etc.) Limited life materials may be life expired, reference materials may be exposed to sunlight which bleaches colour and reduces strength.

- Identity (type numbers, batch numbers etc.) If only one type of steel is procured then marking each piece is unnecessary.

- Warning notices, when relevant, of hazards to product and personnel.

Establishing the root cause

The auditor should not confuse symptoms with causes. A symptom is an observable phenomenon arising from a nonconformity[4], e.g. document not approved, result not meeting requirement, operation not carried out. A cause is a proven reason for the existence of the nonconformity. Approving the document, reworking the product, or carrying out the omitted operation will not eliminate the agent that caused its existence. There is another reason or number of reasons why these nonconformities exist. Among the reasons will be a dominant cause, one which has to be eliminated to prevent recurrence.

When a potential problem is discovered the auditor should establish the cause by asking the auditee to explain why the particular situation has arisen. The answer may prompt a further 'Why?' question and so on until the root cause is established. Along this trail, the auditee may not know all the answers but may be able to direct the auditor to someone who does. The why-why? technique is very effective if used politely and with consideration for the auditee. Asking why, up to seven times, will often reveal the root cause. Some of the causes may appear unimportant, such as 'The manager was on annual leave'. This should prompt a further question: 'Why was no one assigned to carry out the task in his absence?' The trail may reveal several nonconformities, such as in the following scenario.

Having discovered that there was no record of the inspections being carried out, the auditor asks:

[4] Juran, *Quality Control Handbook*, Section 2, Third edition 1974

> *Auditor.* Why is there no record of these inspections?
>
> *Operator.* The procedure does not require the inspections to be recorded.

The auditor asked for sight of the procedure which prescribes the activity and verified that records were not required. Nonconformity No 1 with clause 4.10.1. The records required were not specified in documented procedures.

(The auditor also noticed that the document in use had not been approved. Nonconformity No 2 with clause 4.5.2. Document was not approved prior to issue.) The auditor continues:

> *Auditor.* Why was the procedure not approved before issue?
>
> *Operator.* Because the manager was absent.
>
> *Auditor.* Who is assigned to approve documents in his absence?
>
> *Operator.* Well, er, no one is.
>
> *Auditor.* Why was the unapproved document issued for use?
>
> *Operator.* Because it was needed urgently.

(Nonconformity No 3 with clause 4.1.2.1. No one had been assigned to initiate action to prevent the occurrence of a nonconformity relating to the quality system.)

> *Auditor.* What procedure do you employ in such circumstances?
>
> *Operator.* We don't have a procedure for that.

(Nonconformity No 4 with clause 4.14.3. Procedure not established for determining the steps needed to deal with any problems requiring preventive action.)

The auditor now examines the document control procedure and discovers that no criteria is stated for determining the adequacy of documents. Nonconformity No 5. Document control procedure does not specify criteria which ensures that documents are reviewed and approved for adequacy by authorized personnel prior to issue.

The auditor now examines the operator's instructions.

> *Auditor.* Where is it stated that only approved documents shall be used?
>
> *Operator.* I don't know.
>
> *Guide.* I think that is stated in the Quality Manual.

(The auditor examines the Quality Manual and confirms this to be true.)

Auditor. What instruction have the staff had in using the Quality Manual?

Operator. Oh, we were only trained in the use of our procedures.

Auditor then thanks the operator for his co-operation and asks the guide to take him to the operator's supervisor. After the preliminaries the auditor asks:

Auditor. What instruction have the operators received to familiarize them with the Quality Manual?

Supervisor. It was not considered necessary to familiarize them with the manual as the procedures implement all the policies in the manual.

Auditor. OK I understand. Which procedure implements the policy on restrictin the use of unapproved documents?

Supervisor. You've got me there, I don't think that policy is covered in any procedure.

(Nonconformity No 6 with clause 4.2.2a. Procedures have not been prepared consistent with the requirements of ISO 9001 and the supplier's stated quality policy.)

The auditor sums up the situation with the guide.

Auditor. So had the operators known that they were not supposed to work to unapproved documents the operation may not have commenced; but the cause of the problem seems to be that no procedure has been established to determine the steps needed to deal with any problems requiring preventive action

This was Nonconformity No 4. However, it will be apparent that continuing to question beyond the root cause will still reveal further nonconformities. These were all minor nonconformities except for the root cause which is a major nonconformity, since it proves the system itself is not causing compliance and preventing noncompliance. It is a situation that will recur and one which may cause more significant nonconformities in due course.

Auditor conduct

The essence of good auditors is their ability to reveal the facts swiftly and yet leave the auditee with a feeling that it was a beneficial experience of value to the organization. Auditors can, however, move slowly, uncover little of any consequence and do so in a manner that irritates and annoys the auditee, leaving nothing of any value to the organization. The most important skill an auditor must posses is communication skill: the ability to communicate orally, in writing and with appropriate body language. The auditor needs to leave behind a good impression, that he/she was professional and provided something of value to the organization. Auditors will be remembered for two things, their conduct and their objectivity. They also leave behind two things, an impression and a report, and if the impression is one of a nit picker, their report will not be taken seriously. However, even if they leave behind a professional impression, this may be destroyed by a report which contains irrelevancies. Auditors will only be taken as seriously as their most insignificant nonconformity.

There follows some tips for conducting the audit in various circumstances.

The interview location

- Find a suitable place for the interview away from noisy machines.

- Don't conduct the interview in a conference room as there will be nothing there that you wish to examine.

- If barriers such as desks and tables cannot be avoided, as in small offices, sit alongside the auditee, not opposite.

- Try and keep at the same physical level as the auditee when interviewing. If they stand you stand, if they sit you sit within reason. Standing alongside an auditee sitting at a desk to examine some evidence is more friendly than standing with the desk between the auditor and auditee.

The style of behaviour

- Use a friendly tone, not an aggressive one.

- Always be polite and request information or request to examine something. Do not demand. Whilst the auditor should seek objective evidence don't say 'Show me' without being polite.

- Be assertive but be sensitive to the culture and the position of the person in the hierarchy.

- Control the interview by asking pertinent questions. Remember that the one asking the questions is in control not the one answering.

- Don't be passive and let the auditee control the interview.

- Don't be persuaded to accept evidence that is not conclusive.

- Select your own samples and the people you wish to interview.

Oral communication techniques

- Oral communication should convey the intended meaning clearly.

- Be careful of the tone of speech. The same words spoken with different tone can imply different meaning.

- Enter into a dialogue with the auditee in which awareness and understanding is created.

- Form your questions out of what the auditee says. Do not read from your check list and ignore their responses.

- Be careful with sounds of acknowledgement. A hmm, or ah, ah! in the wrong tone can signal that you have found an error.

- Sound appreciative and positive. Give praise where praise is due. Compliment people on their work if it impresses you.

- Don't criticize work, behaviour of others or the looks of product or documents; it can alienate you and the auditee. Whilst not being up to your standards it may well be adequate for the particular circumstances.

Non-verbal communication

- The facial expression when asking questions can change the intended meaning.

- The facial expression when listening to answers can give the wrong impression.

- Look people in the eye when talking to them but do not stare. Sometimes in the culture it is impolite to look people in the eye.

- Give the impression that you are listening, don't stare around and look out of the window or at what someone else is doing.

- Be aware of cultural differences and customs such as hand gestures and body movements.

Questioning techniques

Explaining your needs

- Give the reasons for asking questions such as:

  'I am seeking to establish that training needs are identified in the organization. Could you tell me how you identify the training needs in your organization?'

  'I am seeking to establish what design proving is conducted before production commences. Could you outline the methods you use?'

> **TYPES OF QUESTIONS**
> - Open questions
> - Closed questions
> - Probing questions
> - Emotive questions
> - Trick questions
> - Leading questions
> - Hypothetical questions
> - Systematic questions
> - Leading questions
> - Multiple questions

- Don't demand information, request it politely. For example:

  Say: *'May I see examples of some recent design review records?'*

  Not: *'Show me your design review records.'*

- Don't use deception to obtain information such as waiting until the auditee has left the office to examine the files.

Steering the auditee

Lead the auditee to the information you want. They may not understand your question so don't get irritated, try an alternative way of saying the same thing, give suggestions by way of illustrating what you want to see. For example:

  You may have said: *'How do you handle nonconforming material?'*

  Rephrase this as: *'How do you deal with items that have failed to meet requirements?'*

  And you can add: *'Let's start with products that you find unacceptable on receipt from your suppliers.'*

Open questions

Ask open questions, questions which require the auditee to explain something. These type of questions commence with one of the following words: what, where, why, when, who and how. For example:

  'What is the company policy on design reviews?'

  'Why do you only perform one design review?'

  'When is the review carried out?'

> **OPEN QUESTIONS**
> - What is the ...?
> - Where do you ...?
> - Why does ...?
> - When is the ...?
> - Who is responsible for ...?
> - How are ...?

'How is the review process controlled?'

'Who attends the design review?'

'Where are the records of these reviews stored?'

'Could you please show me some examples of recent design review records?'

Closed questions

In general don't ask closed questions. For example:

> *'Do you have a procedure for controlling non-conforming product?'* Auditee answers: *'Yes.'*
>
> *'Is the procedure approved?'* Auditee answers: *'Yes.'*
>
> *'Does the procedure cover software?'* Auditee answers: *'No.'*
>
> *'Is the software procedure listed in the Quality Manual?'* Auditee answers: *'No.'*

> **CLOSED QUESTIONS**
>
> - Could I see ...?
> - Have you got ...?
> - Are you responsible for ...?
> - Does this ...?
> - Is the ...?
> - Can you show me ...?
> - Do you have ...?

> *'Can I see a copy of the software procedure?'* Auditee answers: *'No.'*

> *'Why not?'* Auditee answers: *'Because we don't use any software.'*

Closed questions are useful, however, to confirm understanding. For example:

> *'Am I correct in stating that the only customer complaints that are recorded are those received in writing from the customer?'* Auditee answers: *'Yes.'*
>
> *'... and that verbal complaints from customers given directly to your service engineers are not captured by your customer complaints procedure?'* Auditee answers: *'Yes.'*
>
> *'... and your customer complaints procedure does not distinguish between written and verbal complaints?'* Auditee answers: *'Yes, I suppose it is unclear.'*

The auditor has now confirmed that the customer complaints procedures is not effective as it does not handle all customer complaints and therefore fails to meet the requirements of clause 4.14.2.

Emotive questions

Don't ask emotive questions as you might be displaying a prejudice for certain methods. For example:

> *'Don't you use red reject labels?'*

> *'Surely you must use a mater record index?'*

> *'Why don't you stamp these documents UNCONTROLLED?'*

The Tale of an Aborted Opening Meeting

Colin was a competent auditor and there was no doubt that he was assertive. He had met the Managing Director only briefly on the pre-audit visit – a career woman who was proud of her company. However, at the Opening Meeting Colin was not prepared for a fight.

During his training, Colin had learnt not to allow the Opening Meeting to drag on for too long and was determined that this meeting was not going to run beyond 15 minutes, as they had a tight schedule to meet.

Colin and his team were greeted by the MD and her staff at precisely 0900hrs. The auditing was scheduled to commence at 0915 and Colin had some 12 points to cover on his agenda before then.

After Colin had introduced his team he requested Sandra, the MD, to introduce the company personnel present. Following the introductions Colin was quick to regain control and proceeded to explain the scope of the audit. Whilst outlining the audit plan, he was interrupted by Sandra who expressed a desire to make a presentation about the company. Colin politely declined the offer and indicated that he was going to move on. Sandra, not taking this rebuff lightly, stood up and insisted that she gave a presentation as none of the other auditors had been to the site previously.

Colin again declined but more forcefully. Sandra stood her ground and again pressed Colin for time to give the presentation. Unfortunately Colin was paying too much attention to his watch to notice that Sandra had begun to explain some diagrams she had prepared earlier on a flip chart. Colin now felt he was being ignored and this he didn't like. To exert his position he stood up and requested the presentation be deferred until lunch time.

Without further argument, Sandra picked up her papers, beckoned to her staff that the meeting was over and proceeded out of the room. She remarked to Colin as she passed him that she would be in touch with his company concerning his conduct. Colin sat down and wondered where he had gone wrong.

The moral of this story is:

You may be the team leader but you must always behave as a guest!

Trick questions

Trick questions are those designed to trick the auditee into giving an answer that they would not have given if the question had been more open. For example:

'*When did you stop releasing nonconforming product?*' Incorrectly implying that they make a habit of releasing nonconforming product.

'*It is stated in your procedures that all documents have to carry an approval signature of a manager so why have these National Standards not been signed?*' Incorrectly implying that the procedures apply to documents of external origin.

'*I notice you have a clock on the wall in the test laboratory. Where is the calibration status label?*' Incorrectly implying that the clock is a measuring instrument used in testing product.

Leading questions

Don't ask leading questions which contain an assumption since the auditee is unlikely to refute your claim. For example:

'*I expect you check the documents before they are released?*'

'*I presume you send out an agenda before the management review meeting?*'

'*Presumably all purchase orders contain provision for adding pre-qualification requirements?*'

The auditee's answer to all of these questions will be 'Yes'.

If you find yourself asking leading questions then after an answer in the affirmative, the way to recover the situation is to ask: 'Could you show me some examples please?'

Hypothetical questions

Often the information examined may not show evidence that operations would remain under control when unusual circumstances arise. The auditor may therefore have to get the auditee to imagine hypothetical situations in order to verify that the procedures cater for realistic circumstances. The auditee may be able to recall where instances have occurred in the past and the auditor can seek out the evidence. For example:

'*What if the results of the design review indicate that the design needs to be changed, how would these changes be handled?*'

'*What if the customer phones through a change to the contract and requests you implement it immediately?*'

'*What if the supplier fails to send the required paperwork with the consignment?*'

You need to ensure that such hypothetical situations are within the bounds of possibility. Exceeding the bounds of possibility is akin to asking a trick question. However, be wary of the answer 'It never happens'. The standard requires a procedure for handling customer complaints. Some suppliers do not receive complaints from their customers. This is no reason not to have a mechanism for determining the level of customer satisfaction. Ideally the supplier should have carried out a failure modes analysis to detect ways in which the quality system may fail to prevent nonconformity. Whilst not a requirement of ISO 9000, a quality system which does not cause conformity and prevent nonconformity is not effective, therefore the auditor should verify that the system will prevent potential failures.

Systematic questions

These are the most common questions to ask when following an audit trail. They are simply in the form: 'What happens next?' or 'OK, you send out the design review agenda, what happens next?'

Multiple questions

Avoid asking multiple questions. They tend to confuse the auditee and often are not completely answered as the conversation will diverge onto other subjects before all answers are given. The auditee then forgets what you originally said. More importantly, you may forget what you asked, which shows an undisciplined approach. For example:

> 'Which of these products was reworked, where is the inspection record and what happened to the others which passed inspection?'

> 'Could you explain the meaning of these qualifications and let me see a copy of this person's training records and his job description and competency requirement?'

Requests

In addition to asking questions, auditors need to examine objective evidence. To obtain sight of objective evidence auditors need to request that the auditee carry out an action rather than provide an answer. The answer to the question 'Can you please show me a copy of the customer complaints procedure?' is either a 'Yes' or a 'No' and doesn't get the auditor any further in his/her quest for information. If you are faced with an unco-operative auditee then this may well be the response. Auditors therefore need to express their request as an instruction. For example:

> 'I would like to see some results of the design verification activities you performed on Project Nexus.'

> 'Having examined your audit procedures and found them satisfactory, I would now like to examine the file of audit reports.'

> 'I would like to look at the training records of your auditors if you have them available.'

The Lead Auditor's role during the audit

Depending on the size of the team, the time which the Lead Auditor spends auditing and managing the team will vary. It is normal practice for the Lead Auditor to audit the areas of business management and quality system management and for this a guide may not be necessary as the people concerned may reside at one location. After the Opening Meeting the Lead Auditor should ensure each member of the team has made contact with the appropriate guide and is being taken to their first appointment. During the audit, the Lead Auditor should:

● Periodically check on progress with each of the auditors

● Determine if the auditors are having problems with the guides

● Determine if any major nonconformities have been detected

● Take up with the Management Representative any issues that may need to be resolved

Giving feedback

When gathering information on the quality system, the auditor should indicate either verbally or using body language whether what has been offered is acceptable. Nodding the head in agreement, polite interaction in the dialogue are acceptable forms of feedback. In addition:

● When something is found that is in your opinion good, give praise.

● When problems are uncovered, don't jump to conclusions.

● Don't indicate that there is a nonconformity with the standard as there may not be. Other facts yet to be revealed may negate a nonconformity.

● Point out potential problems, inadvertent errors and mistakes and suggest that the auditee might like to initiate their removal. Do this verbally and politely being sensitive to the feelings of others in the vicinity.

● If there is evidence of a nonconformity in the system then say that there is a weakness not a nonconformity, as the very term nonconformity is a demotivator.

Giving advice

It is common practice for third party auditors not to give advice as to how a nonconformity may be corrected as it may result in a conflict of interest. A conflict of interest will only arise if the auditor advises a particular solution which would incur additional cost or which is ill-considered and may impact health, safety or working relationships. This is a legal issue rather than a commercial one as he who gives advice must take responsibility for the consequences. Also one auditor's solutions may conflict with another auditor's solutions. A different auditor may find new nonconformities which were present during the previous audit. This can be claimed as being the consequence of sampling. However, if the company has implemented a particular solution suggested by an auditor, and this having been implemented exactly as advised results in a subsequent auditor reporting the practice is in nonconformity with the standard, then the company has just cause for complaint to the certification body. As certification bodies don't like receiving complaints about their auditors, they tend to prohibit their auditors from giving advice. The rules are as follows:

- Don't provide specific solutions to nonconformities.

- If asked, either refuse or indicate a range of possible solutions that would be acceptable from which the auditee can decide which would be appropriate in their circumstances.

- Don't refer to any particular company that employs any methods suggested.

- Don't indicate that a particular solution they propose would not work. Refer always to the requirements of the standard.

- Don't say that the company should do various things, phrase suggestions as opportunities for improvement. For example: 'The frequency of errors may be reduced by providing staff with the reasons for some of the more important instructions.'

Auditee conduct

The conduct of the auditee is crucial to the effectiveness of the audit and if the auditor is not assertive it can significantly reduce the value of the audit results. Most of the time auditees are responsive and co-operative as they desire to create a good impression. However, not everyone in an organization will feel

AUDITEE CONDUCT
• Wasting time
• Provocation
• Insincerity
• Language
• Bribes

the same way and some may regard the exercise as negative and with their inside knowledge of the organization may attempt to sabotage the audit either by painting a rosy picture when in fact the opposite is true, or by delaying tactics. Here are some which auditors should know how to detect and deal with effectively.

Wasting time

Auditees may waste time by late arrival, long lunches, protracted discussions on principles or about nothing in particular, formal presentations, a tour of the site, forgetting requests deliberately and many other similar tactics.

The auditor should be alert to such practices but should not jump to conclusions. The behaviour may be cultural or depending on the position of the auditee, some respect needs to be given to their ego, successes, pride etc. Before taking any action the auditor should be satisfied that the behaviour is devious and not inadvertent. If inadvertent, then a polite request to continue with the audit or to delay the presentation or tour until a more appropriate time may suffice. If in fact delays happen often, then the auditor has to take action and suggest that unless progress is made more quickly, the audit will have to be extended as the time allowed did not include provision for this situation.

Provocation

Sometimes an auditee may be just as knowledgeable about ISO 9000 as the auditor. This is especially the case when interviewing the Management Representative. The auditor needs to remember that the Management Representative may have developed the system, have been on the training courses and in fact may be a registered Lead Auditor. In such cases the auditor needs to be on his/her guard.

The only way to combat such a situation is to be sure of your facts before they are declared and use disarming techniques such as acknowledging the person's experience, respecting their position and experience rather than entering a contest as to who is right. It is not a question as to who is right but what is right and if you can prove by objective evidence that a specific requirement of the standard had not been met then you have a case and in most instances will convince another professional of the situation. If you go beyond the requirements of the standard and claim the company is noncompliant with non-existent requirements then you deserve all you get.

Insincerity

This can arise either due to flattery or pity. Sometimes the auditee will be over complimentary in their praise, their appreciation and their gratitude. On other occasions they may play on your kindness and expect favours in view of the consequences of a failure to be certified. The company could go out of business, staff could be laid off, the quality manager may lose his / her job as their tenure depended on the company receiving an ISO 9000 certificate.

The only action an auditor can take in such circumstances is to ignore such pleas as irrelevant to the results. Be polite, express sympathy but as your job is to reveal facts which were already present in the organization, you have a duty to report as you find and being a professional, you are sure that the company would expect no less of you. If you were to compromise your findings on the basis of the consequences, suggest that the company may not like such considerations to be given to their competitors and that the only way to conduct such audits is on an equitable basis.

Language

It is essential that both auditee and auditor are fluent in the language in which the audit is to be conducted. If there are language difficulties then an interpreter should be selected by the auditor to translate all questions and answers. If the auditee provides the interpreter one can never be sure that the translations were accurate.

Bribes

Auditors may be offered bribes to ignore findings or fudge the results, as a contract may depend on the result. A good lunch or a gift of a pen or other articles which are often offered to customers should be accepted gratefully if offered after disclosure of the results, as to refuse may offend, but no favours should be given as a result. They expect a fair and professional assessment. If gifts are offered before the results are disclosed, decline them if intended as a bribe.

Dealing with challenges

Often an auditor will ask a perfectly sensible question and be challenged by the auditee. Sometimes this is a defensive attitude or it could be that the auditee wishes to learn from the auditor. In response to your question the auditee may say for example:

Do we have to do that?

Sometimes this is the response to a question such as: 'Where have you defined your equipment maintenance requirements?' The standard does not require that equipment maintenance requirements are defined but it

> ## *CHALLENGES*
> - Do we have to do that?
> - Where does it say that in the standard?
> - Why do we need to do that?
> - We don't believe the standard requires that.
> - Our interpretation is different.
> - We checked that on internal audit.

does require that equipment be maintained in clause 4.9g. The auditor should not answer 'No' but restate the question. If this gets the same response, then rephrase the question such as 'What provisions have you made for equipment maintenance?' If this gets the same response then explain that you are seeking information to establish that the quality system has been established and documented (a requirement of clause 4.2.1 of the standard) and that the question is not intended to prove nonconformity but conformity with the requirements of the standard.

Where does it say that in the standard?

This response is similar to the one above but more specific. For example, you may have queried that there are no records of equipment maintenance in which case your expectations are beyond what the standard actually requires. There is no specific requirement for equipment maintenance records, only a requirement for suitable maintenance to ensure continuing process capability. Instead of showing your ignorance by attempting to find the requirement in the standard, rephrase the question. Say, 'Let me rephrase the question. How do you maintain equipment to ensure continued process capability?' And then, examine the response.

Why do we need to do that?

This response may arise if you indicate in your question a particular solution to a requirement, such as expecting documents to carry an approval signature or a change record. You need to explain the reason for your question by relating it to the requirements of the standard. If you inadvertently suggest a solution in your question, rephrase the question by explaining that many companies denote document approval by signature on the document or include a change record to indicate the nature of change, and such a practice is an acceptable way of meeting these requirements. Then ask how they signify that documents have been approved and the nature of change identified.

We don't believe the standard requires us to do that

You may have implied that the standard requires a documented procedure for management review by asking to examine the management review procedure. Again, don't make assumptions or show your ignorance by consulting the standard in front of the auditee, as the standard does not require a documented procedure for management review. Rephrase the question, such as: 'In what document do you describe how you conduct your management reviews. If this gets the same response, then you can explain that in clause 4.2.1 of the standard, it requires that the quality system be documented and that the Quality Manual cover the requirements of the standard. Since management review is a requirement of clause 4.1.3, it follows that a document should exist in the quality system which defines how the management review requirements are met. So you can ask: 'In what document have you defined your management review practices?'

We have not interpreted the standard as you do

This response can cause argument if your interpretation is not based on sound evidence. For example, you may have asked to examine their List of Approved Suppliers, and they show you records of suppliers they use but you find that they have used a supplier not listed in the records. If you conclude that they have used an un-approved supplier they may challenge your claim with this response: 'The standard does not require that suppliers are approved, neither does it require suppliers to maintain a list of approved suppliers.' Only if their defined policy is that they will only select suppliers from a list of approved suppliers can you conclude that they are not conforming to the standard. They are required to select their suppliers on their ability to meet subcontract requirements and also define the extent of control exercised over subcontractors, which means that they can use unacceptable suppliers providing they exercise adequate control. If it can be demonstrated that adequate control has been exercised over the supplier that is not identified in their records of acceptable suppliers, then there is no nonconformity.

That was checked on the internal audit

This response is likely if you have asked to see evidence of an operation or decision that appears to be no longer available. Before defending your position, you need to be sure that the standard requires what you are asking for. The standard does not require the organization to record everything, neither does it require records to be maintained to demonstrate that every one of its requirements have been met. Producing records and maintaining records are two different things. For example the standard does not require either proposed or implemented corrective actions to be recorded. It only requires the results of the investigation into the cause of nonconformities to be recorded. Therefore if you have asked to see the corrective action records and they have not been kept, then tough. You will have to find another way of verifying the requirement. There is a simple solution. Ask to see the records of the internal audit which examined corrective action procedures, or if challenged by the response: 'That was checked on the internal audit', ask to see the reports. If the reports show that the corrective action procedures were followed then unless you have reason to suppose they are fabricated, you should accept them. Remember, try to find a reason to pass the company rather than a reason to fail them.

Recording results

Audit notes

The results of the audit need to be recorded as the audit progresses so as to provide the data that will be used later to compile the audit report. As facts are established, the auditor should record:

- The facts that indicate conformance

- The facts that indicate a possible nonconformance

- Observations on effective practices

- Observations on ineffective practices

- Examples of impressive performance, products, documentation, practices, conditions, attitudes etc.

In noting facts, details such as the identity of products, documents, locations etc. should be recorded. Some samples of audit notes follow:

- Procedure QA 005 defines annual management review intervals. (Facts indicating conformity)

- Management review records for June 93 and January 95 don't indicate conclusions and recommendations for corrective action. (Facts indicating possible nonconformity)

- Audit of quality department performed in June 94 carried out by quality engineer. (Facts indicating possible nonconformity)

- Quality objectives relevant to organizational goals have not been documented. (Facts indicating possible nonconformity)

- 3 out of 5 inspectors in assembly shop inspecting printed circuit boards trained by watching another person and claimed no knowledge of PCB inspection procedure QC 034. (Facts indicating possible nonconformity)

- A supplier of item KL 89756 had been selected on the basis of price by the purchasing manager – no record of Crump valves performance. (Facts indicating possible nonconformity)

- No evidence that unauthorized receipt into and dispatch from raw material stores would occur. (Observation on effective practices)

- No evidence was found for customer complaints serial numbers CC93/001 and CC93/002 reporting receipt of defective power supply model PS 54783G to show that corrective actions had been taken to prevent their recurrence. (Observation on ineffective practices)

- Level of rejected components shown in production histograms for 1994 reduced from 100 per day to 1 per week over the period of two months. (Examples of impressive performance)

- Good correlation between policies and procedures in Quality Manual. (Example of impressive performance)

These statements are all facts but not necessarily nonconformities. Some indicate conformity and some are observations. (See Chapter 5 for an analysis.)

Agreeing the facts

The findings resulting from an interview or series of interviews in an area should be discussed with the guide before moving on to the next area. In this way the opportunity to remove any misunderstanding can be taken whilst on the scene of the action, thereby avoiding the need to return later. When the auditor enters an area or commences an interview he/she should have identified their objective, i.e. what they want to establish. Before leaving the area, the auditor should establish that they have gathered sufficient data to draw a conclusion that their objective has been achieved. Agreement to the facts is not agreement to nonconformity, as other facts may emerge later which alter the conclusion. So when the guide is asked to confirm certain facts, they are being asked to confirm what was found and not whether there was a nonconformity. For example, no records were found to show that a certain product had been inspected, or no evidence could be found that a purchase order had been reviewed and approved before release, or a certain piece of test equipment was in use after the indicated due date for calibration. Agreement at this stage is verbal. Later, when the formal corrective action report has been produced, the guide can be asked to sign the statement as a true fact if no further evidence has been found to confirm compliance.

The auditors' review meeting

During the audit, each auditor will collect objective evidence on the performance of the organization's quality system. At the discretion of the Lead Auditor, team review meetings may be held prior to lunch and prior to reporting the results to the company at the end of each day. These meetings should be a forum where:

- Problems can be discussed with other team members.

- The audit plan can be revised if areas not in the original plan are uncovered or if the timing requires adjustment.

- The Lead Auditor can direct team members to look for certain evidence or focus on particular aspects which need closer attention.

- Check lists can be reviewed and modified if necessary.

- It can be established that the audit is probing the quality system sufficiently to verify conformity with the requirements of the standard.

- Team members can exchange information and leads for others to follow in securing objective evidence.

- Interpretations of the standard can be harmonized.

- Audit findings can be discussed and the views of other auditors considered in determining if there are nonconformities.

- Audit findings can be analysed to determine if they are common to more than one area.

- Corrective action reports can be prepared (see Chapter 5).

- The Lead Auditor can determine if the Management Representative needs to be informed of issues that require his/her attention.

The auditors' review meeting is optional but with a team of more than two auditors is often necessary to ensure a consistent approach and avoid problems which may jeopardize success.

Summary

This chapter addressed the process of conducting the audit and showed how important it is for auditors to remain calm, cool and objective and concentrate on gathering facts rather than faults. The following are the key issues covered:

◻ The documentation audit is a key element of the audit and is carried out prior to the Opening Meeting either off-site or on-site.

◻ The documentation audit requires examination of all quality system documents since requirements of the standard may be addressed at any level in the hierarchy.

◻ There are over 300 requirements in ISO 9001 so the documentation audit is not a simple task.

◻ The results of the documentation audit should be conveyed to the company and sufficient time allowed for remedial action. Such action may involve changes to the documentation but may also require the implementation of new practices.

◻ Opening Meetings are convened to confirm the arrangements for the implementation audit.

◻ Opening Meetings should be kept short, no more than 15 minutes.

◻ Confirmation of the purpose and scope of the audit is the key to understanding the direction of the audit and the success criteria needs to be defined at the outset.

◻ Auditors seek conformity and come across nonconformity and not the other way around.

◻ The audit is limited to the scope of the standard and products and services supplied by the company.

◻ Obtaining information depends on seeking out the right person and knowing what you want to establish, not going in like an unguided missile.

◻ In establishing conformity and nonconformity it is important to take several samples that are representative.

◻ The best way to conduct an audit is by following a trail rather than checking against the clauses of the standard in the sequence they are presented.

◻ Obtaining information is a matter of asking opening questions and confirming facts with closing questions in the presence of a guide who can witness the event.

◻ Use the why-why? technique to reveal the root cause of the nonconformity.

◻ Auditors have to be observant to discover conformity and nonconformity, explain their needs and steer the auditor to their goal.

◻ Auditees can sabotage your efforts, so knowing the counter measures to take is important.

◻ Auditors have to remain in control but also remain polite, professional and assertive.

◻ Communication skills are the most important of all the skills an auditor should possess.

◻ Questioning techniques need to be learnt to gain knowledge about the process through people, as auditors audit the process, not the people.

◻ Feedback is important to demonstrate the information that is given is needed.

◻ Giving specific advice is prohibited but that doesn't mean the auditor can't seek improvements.

◻ Knowing how to deal with challenges effectively puts the auditor in control and demonstrates your capabilities.

◻ The presence of guides and note-taking is valuable for compiling a factual report that will sustain challenges.

◻ The Lead Auditor has a key role in maintaining the schedule, overcoming problems and conducting auditor review meetings to secure team agreement to the findings.

Chapter 5

Reporting audits

The reporting process

The audit reporting process commences by determining which findings are to be reported and continues through to the delivery of the report with conclusions at a Closing Meeting. This process is illustrated in Figure 5-1.

Analysing the evidence

What to report

Auditors are seeking evidence of conformity with the requirements of the standard but their aim is to establish whether the quality system is effective. These objectives may appear to conflict as a failure to meet a requirement of the standard does not necessarily indicate that the quality system is ineffective. If one document in use had not been approved prior to issue, is the system ineffective? If one person was not wearing the prescribed protective clothing, is the system ineffective? It depends on whether there are many documents in use or only one and whether there are many people who need to wear such clothing or only one and what the consequences are. It also depends upon what

WHAT TO REPORT
● Good points
● Degree of conformity
● Degree of nonconformity
● Improvements
● System effectiveness
● Conclusions

the cause of the nonconformity was. The auditor should attempt to establish the cause by using the why-why? technique as explained in Chapter 4. If the root cause has been

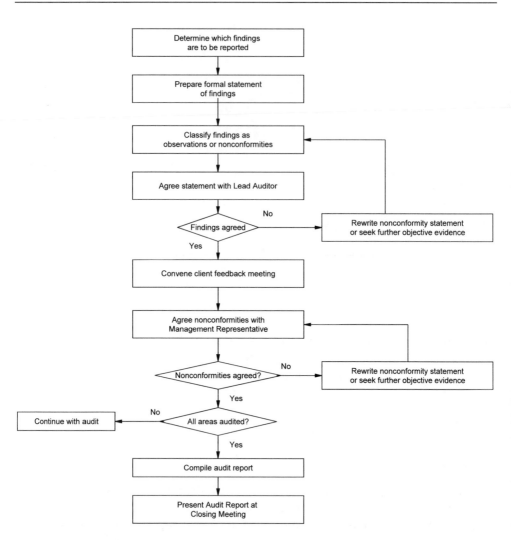

Figure 5-1 Audit reporting process

established then its resolution should prevent the string of nonconformities found en-route. In such cases it may be prudent for the nonconformity statement to address only the root cause nonconformity and omit the others. All the findings therefore do not need to be reported, only those which demonstrate that the quality system is either effective or ineffective. In simple terms effectiveness is a measure of achieving two objectives (see later for a more detailed analysis): Will the system cause conforming product/service to be supplied and prevent nonconforming product/service from being supplied? If the findings cannot be shown to serve either of these two objectives then there is little value in requiring corrective action. In the example scenario given in Chapter 4, it will be apparent that approving the procedure would not in fact prevent recurrence of the problem and that a more fundamental problem existed which needs to be addressed.

The company needs to be informed about the extent of compliance with the requirements but in dealing with any nonconformity, they need to be informed of those areas where they need to take corrective action. Corrective action is action taken to prevent the recurrence of a nonconformity. If no action can be taken to prevent recurrence then no action should be requested by the auditor, although indication of where improvements could be made should be offered if they add value. If correcting the nonconformity does not improve the quality system then the incident is isolated. Isolated incidents which can be corrected on the spot are not nonconformities. They are inadvertent errors (see Chapter 3 on *Success Criteria*).

When to request corrective action

Reporting of nonconformities is a request for corrective action. Corrective action should therefore be requested:

- When evidence of frequent errors has been found

- When there is evidence that the quality of all products or services of a particular type will be, is being or has been adversely affected

- When there is objective evidence that the quality system will not cause conforming product to be delivered

- When there is evidence that the quality system will not prevent delivery of nonconforming product or service

- When there is evidence that the quality system will not cause objective evidence to be generated to demonstrate that operations affecting quality are under control

and where all this evidence can be traced to a requirement of the standard.

When not to request corrective action

- When an isolated error can be corrected on the spot

- When errors are attributable to the gestation period for the quality system

- When errors are attributable to new staff being on the learning curve

- When an individual knows of inadvertent errors and accepts remedial action

- When known problems are being resolved through an approved corrective action plan

- When practice cannot be proven to be at variance with the requirements of the standard

Company feedback meetings

When an audit is planned to take more than one day, company feedback meetings should be held to:

- Report progress against plan

- Request extension to the programme if delays beyond the audit team's control have occurred

- Request changes to the plan to include areas not identified in the original plan

- Seek clarification of the company's quality system, its products and services and the means by which it ensures their quality

- Report problems in executing the audit that require action by the company

- Request information needed to complete investigations

- Report nonconformities and observations

OPTIONS FOR AGREEING NONCONFORMITIES

- With auditee at the time they are found (Aggressive behaviour)

- With guide prior to leaving the area (Formal statement difficult to prepare on the spot)

- With company management at daily company feedback meetings (Recommended)

- With Management Representative prior to Closing Meeting (Recommended for 1-day audits)

- With company management at Closing Meeting (Can be traumatic)

The company feedback meetings are of mutual benefit to the audit team and the company but should not degenerate into a confrontation. It is therefore important that the first feedback meeting be taken as an opportunity to show that audits add value. So no nit picking. In fact only report the significant nonconformities and observations. A first feedback meeting at which trivial incidents are reported may give the impression that the team are unprofessional.

The company feedback meeting can be held at the end or at the beginning of the day following an auditing day.

The advantages of same-day company feedback meeting:

• The findings are still fresh in people's minds.

• The guides and managers of the areas audited are still available.

• Limits the time spent in the meeting as people will want to get home.

The disadvantages of same-day company feedback meeting:

• Reduces auditing time as audit team need a review meeting beforehand.

• Reduces potential for company to take corrective action on the same day.

The advantages of next-day company feedback meeting:

• Audit team review meeting can be held in the evening rather that during the day and hence provides for more auditing time.

• Company can begin to take corrective action immediately.

The disadvantages of next-day company feedback meeting:

• Some managers may have other commitments.

• On-site audit team reviews may be avoided altogether and lose their other advantages.

• Could eat into auditing time unless well controlled as company will want to debate the issues.

Completed nonconformity reports may be presented for the agreement of the Management Representative. However, it is desirable to avoid leaving copies of the reports with the company as facts may be revealed later which warrant a change to the report or its withdrawal. If copies of nonconformity reports are handed over, then action may be initiated which is subsequently found unnecessary, a situation which may irritate the auditee. Minor problems may turn out to be major problems as further evidence is revealed, a situation which may confuse the auditee.

Documenting the audit findings

The audit findings are facts revealed during the audit and can be classified as conformities, nonconformities and observations. Conformities should be reported as a check mark against a requirement of the standard rather than a written statement. Nonconformities and observations should be reported by a written statement of fact which includes the objective evidence. The difference between nonconformities and observations is as follows.

● Nonconformities are facts supported by objective evidence which prove a failure to meet a requirement of the standard.

● Observations are facts supported by objective evidence which do not prove a failure to meet a requirement of the standard but which impair the effectiveness of the quality system.

Both types of statements should be written in the same way, in that they should exhibit certain characteristics. Such statements should define the exact incident and its location where appropriate.

The decision tree for determining reportable findings is illustrated in Figure 5.2.

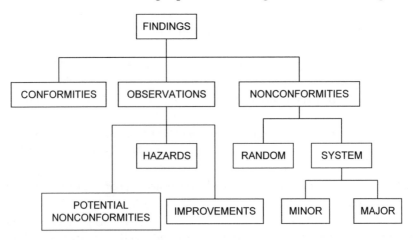

Figure 5-2 Findings decision tree

Whether the findings are reported as observations or nonconformities is only of importance in securing corrective action and making an assessment of system effectiveness. A company committed to quality improvement won't quarrel with taking action on both nonconformities and observations since they both serve to improve the effectiveness of the system. Companies that are not so committed may well not choose to action the observations as there may well be more important issues to deal with. By categorizing findings in this way, one can avoid having to defend a finding for which the objective evidence or the requirement is less than certain. When a finding is categorized as an observation rather than a nonconformity, companies are often quite prepared to accept it. When a finding is categorized as a nonconformity, the implication is that they have made a significant error and therefore it may well be challenged if they feel at all doubtful about its certainty.

Writing the nonconformity statements

Nonconformity statements are findings which disclose the requirement which the organization has failed to meet. Each nonconformity statement should consist of four characteristics:

- The subject of the nonconformity defined in precise terms

- The location where the nonconformity was detected, if appropriate

- The incident which signifies the subject is nonconforming, where relevant

- The requirement from the standard which has not been met

If the subject is not defined accurately then it will not facilitate its location by others. Those within the organization need to take corrective action and unless specific details are recorded they will be unable to do so.

There are several ways in which a nonconformity statement can be written. Some auditors prefer to describe the incident as a story e.g. 'In discussions with the IME Project Manager it was found that the development plan had not been updated since launching the project etc.' There is no need for the preamble. Who the auditor was talking to is irrelevant. What is relevant is the fact the development plan had not been updated. Another example is where auditors use emotive language to describe the nonconformity e.g. 'In contravention of clause 4.4.2 of ISO 9001, the development plan for project IME had not been updated since launching the project'; or another often-used term, 'In violation of clause 4.4.2 of ISO 9001 etc.' There is no need for such language to be used. The emotive terms can irritate company management and imply they have made gross errors of judgement when all that was wrong was that some document had not been maintained. If one looked further, one may find that it had

had no adverse effect on the project since project meeting minutes contained the revisions.

A less emotive way of reporting nonconformities is to be factual; for example, 'The IME Project development plan dated 26 July 1992 had not been updated to reflect changes in the contract statement of work dated 12 December 1993 as required by ISO 9001 clause 4.4.2.' This statement contains the evidence for the nonconformity whereas the others do not.

Another way of preventing alarm is to avoid being global in the nonconformity statements. One might write that 'There were no calibration records for DVM S/N 367183 found in the assembly test area.' There may have been such records. The fact is that none could be found so why not say so:

WORDS AND PHRASES TO AVOID
• In violation of ...
• In contravention of ...
• Whilst auditing xyz I found ...
• Several items of ...
• None of the ...
• Little evidence ...
• Few records ...
• Mr X said that ...
• I observed that ...

e.g. 'No calibration records could be found for DVM S/N 367183 being used in the assembly test area' or in another example 'No evidence could be found to show that issue 5 of the IME development plan had been approved prior to issue.' In this way it indicates that during the audit a search was carried out and the documents required to verify conformity could not be located. Later on these records may be located so the statement can be withdrawn.

Valid nonconformity statements

In the examples in Chapter 4 there is a finding regarding improperly trained inspectors. The facts which were recorded were as follows:

> '3 out of 5 inspectors in assembly shop inspecting printed circuit boards trained by watching another person and claimed no knowledge of PCB inspection procedure QC 034.'

By limiting the nonconformity statement to the bare facts it may read as follows:

> 'Inspectors had not been properly trained.'

Is it a random failure or a systematic failure? Assuming that there were 20 inspectors on the PCB line and a system of training had been developed as verified during the documentation audit, then if the sample size checked was 5 and only 1 was improperly trained, it is a random failure. In the example, further checks to indicate that the sample was representative revealed 3 out of 5 were improperly trained, indicating a systematic failure and therefore a reportable nonconformity. It is not a mistake – the training system has broken down. Corrective action can be taken to prevent recurrence on the PCB line.

Limiting a nonconformity statement to the simple sentence above is totally inadequate. The statement has to define how many inspectors, what they were doing and where they were doing it, for them to be traced and corrective action taken. The statement should be rewritten as follows:

> *'3 out of 5 inspectors in the assembly shop performing final inspection had been trained to do the inspections by watching another person do it.'*

If many types of assemblies are inspected in the assembly shop in which there are 20 inspectors, then to locate the incident, more details are required; so the statement becomes:

> *'3 out of 5 inspectors in the assembly shop performing final inspection on completed printed circuit boards had been trained to do the inspections by watching another person do it.'*

By not specifying the incident, the auditee will be at a loss to understand why the subject was nonconforming. In the example above the auditee may claim that on-the-job training is normal practice so why is it a nonconformity? It is a nonconformity because the inspectors should have been trained to perform the inspection in accordance with PCB procedure QC 034. The statement should therefore read:

> *'3 out of 5 inspectors in the assembly shop performing final inspection on completed printed circuit boards had been trained to do the inspection by watching another person do it and were unaware of PCB procedure QC 034.'*

Disclosure of the incident is not always needed, as another example from chapter 4 shows. 'Management review records for June 93 and January 95 do not indicate conclusions and recommendations for corrective action.' It is not that a decision has been made which conflicts with a requirement but that no decision had been made at all.

In the inspection example above, the auditee again may challenge the auditor, asking where in the standard does it require them to train people to follow procedures. The finding does not yet indicate what is nonconforming. The standard does require that personnel be qualified on the basis of appropriate training in clause 4.18 so the finding can be modified as follows:

> *'3 out of 5 inspectors in the assembly shop, performing final inspection on completed printed circuit boards, had been trained to do the inspection by watching another person do it and were unaware of PCB procedure QC 034. The inspectors were therefore not qualified on the basis of appropriate training by being trained to perform the inspection in accordance with PCB procedure QC 034 as required by clause 4.18 of ISO 9001.'*

In this final version of the nonconformity statement there is no doubt what the subject is, where it is located, the nature of the incident, why it is nonconforming, what specific requirement of the standard has not been met. This is the recommended format of all nonconformity statements. If a clause of the standard cannot be located to prove a nonconformity, there is no nonconformity. Although the sentence is rather long we are not concerned with a work of literature but with a legal statement that will sustain

CONTENT OF NONCONFORMITY STATEMENTS
• The subject
• The location
• The incident
• The requirement

being challenged. However, do try to avoid overlong sentences since they may have to be read out at the Closing Meeting and nothing reduces one's credibility more than stumbling over words in a verbal report.

Other examples in Chapter 3 which are valid when the recorded findings are converted into nonconformity statements are as follows:

- The supplier of item KL 89756 (Crump Valves Ltd.) had been selected on the basis of price and not on the basis of the supplier's ability to meet subcontract requirements as required by clause 4.6.2 of ISO 9001, as no records of the supplier's performance were found.

- No evidence could be found that objectives for quality have been defined and documented as required by clause 4.1.1 of ISO 9001.

Invalid nonconformity statements

An invalid nonconformity statement is one that cannot be validated against the requirements of the standard. The examples in Chapter 4 include two findings that would be invalid nonconformities.

'Management review records for June 93 and January 95 don't indicate conclusions and recommendations for corrective action.' This is not a nonconformity with clause 4.1.3 of ISO 9001 as the clause does not contain requirements for the content of management records. It could be reported as an observation; see below under *Observations*.

In the second example there is insufficient evidence of a nonconformity with clause 4.17. The statement was as follows: 'Audit of quality department performed in June 94 carried out by quality engineer'. The clause requires auditors to be independent of the activities audited but it is not disclosed whether the quality engineer audited activities for which he/she was responsible. Further checks need to be made of the audit records to verify whether he/she was independent and whether the audit procedure requires independent auditors. Some auditors take the view that if the auditor works in the same department then he/she cannot be independent no matter what the responsibilities of the auditor are. Since the auditor's manager is responsible for all activities in the department he/she can influence the auditor's judgement. If we apply this logic through the hierarchy, then as the CEO is responsible for the whole organization he/she can also exert influence on the internal auditor.

Dealing with deviations from company procedures

Provisions beyond the scope of the standard

A quality system and the associated documented procedures may contain provisions that meet the minimum requirements of the standard or may go beyond these requirements. A provision that goes beyond the standard is one which is outside the scope of the standard, such as accounting, credit control, personnel safety, personnel security procedures etc. where these functions are incidental to the products and services provided by the organization. There are no requirements that relate to these provisions unless the product or service of the organization is accountancy, credit control, security etc. Where activities are found to be nonconforming with provisions outside the scope of the standard they should be ignored. Further, the auditor should not be examining such functions during the audit even if requested to do so by the organization. However, where such provisions are required by customers through a contract they should be included.

Specific solutions to requirements of the standard

The procedures may specify certain solutions to the requirements of the standard which are not specifically defined by the standard. Examples may be:

- Approved Suppliers List as a means of selecting suppliers of assessed capability

- Job Descriptions as a means of defining responsibilities and authority

- Statistical Process Control as a means of monitoring process capability

- Quality Costing as a means of determining system effectiveness

In these cases the organization has chosen particular solutions. If the documented procedures provide alternative techniques or options in place of the preferred method and these alternate methods are used, then there is no nonconformity. If there are no specified alternatives or options and the practices are found not to be in conformity with the documented procedures then a nonconformity statement is justified.

An organization may specify particular characteristics, verifications, cleanliness conditions, materials, processes etc. which serve to provide products or services that meet customer needs and expectations. Any failure to achieve or use such measures is a nonconformity unless alternative measures have been approved by authorized personnel and such approval can be verified. Examples include:

- Cardboard packaging used instead of the prescribed expanded polystyrene

- 95KU viscosity achieved when limits are 60-90KU

- Final test for insulation resistance not conducted as specified in test specification

However, there are circumstances when such situations would not warrant a nonconformity. When a specified piece of equipment, material, component etc. is unavailable then in some circumstances no alternatives need be specified in the procedures and specifications if no particular characteristics constrain use. Selection of measuring instruments, tools, techniques etc. is often decided by the trained personnel. If the organization only purchases certain items so that no others are available then choosing from available material may be appropriate. It is the result which is important rather than the means of obtaining it.

Deviations from policies, specifications and procedures can be authorized through approved concessions, waivers, variations etc. that are documented. Providing the person deviating from policy, specification or procedure can show authorization in such form, then nonconformities are not warranted. However, if the concession or waiver authorizes a departure from the standard against which the audit is being conducted then there is a nonconformity.

Provisions within the scope but not prescribed by the standard

The organization's procedures may include provisions which relate to the requirements of the standard but which go beyond these provisions. Examples may be:

- Audit check lists as an aide-memoire to auditors

- Comment sheets used to obtain comments on draft documents

- Acknowledgement slips used for confirming receipt and implementation of procedures

Any failure to implement such provisions is not a nonconformity with the standard unless no other means has been used to carry out the tasks. In the first case, the audit check lists may be the only specified way to plan and audit. If not done, the audit cannot be said to have been planned as required by clause 4.17. In the second case, the comment sheets may be the only way to review documents and if not done the review has not been carried out as required by clause 4.5.2. In the last case the slips may be a secondary measure to ensure the pertinent documents are available at all points of use. This is not a nonconformity unless the documents did not arrive and operations that need it are being carried out. Many organizations use belt and braces to ensure conformity. If the belt fails the braces still hold up the trousers! There is no nonconformity when it is found that personnel are not wearing belts unless of course they are not wearing braces either. In which case the trousers will be found around the ankles!

Auditors need to be seen to be fair and not to insist that all safeguards are adhered to. It is only when the safeguards break down and cause or could cause a product or service nonconformity that some of the measures need to be in place.

Observation statements

Observation statements should be written in a similar manner to nonconformity statements. The difference between the two is that observations are objective evidence which do not prove a failure to meet a requirement of the standard but which impair or may impair the effectiveness of the quality system. Such observations may be situations where:

- There is insufficient evidence of a nonconformity; however, system effectiveness is impaired.

- There is a potential nonconformity.

- Prevailing occupational health, safety or environmental conditions may impact upon product/service quality.

In the examples in Chapter 4 there were findings regarding customer complaints:

'No evidence was found for customer complaints serial numbers CC93/001 and CC93/002 to show that corrective actions had been taken to prevent their recurrence.'

In this example it might be expected that there is a nonconformity with clause 4.14.2 requiring corrective actions to be determined. However, the standard does not require corrective actions to be documented, only the cause of the nonconformities to be recorded and the effectiveness of the actions to be ensured. If further analysis of customer complaints do not reveal a recurrence of defects with power supply model PS 54783G and checks of the corrective action procedure and records of an investigation reveal that they meet the other requirements of clause 4.14.2, then an observation statement is appropriate of the following form:

'*No evidence was found for customer complaints serial numbers CC93/001 and CC93/002 reporting receipt of defective power supply model PS 54783G to show what corrective actions had been taken to prevent their recurrence even though no further complaints had been received to date. It could not therefore be verified that any corrective action that was taken would prevent recurrence as required by clause 4.14.2 of ISO 9001.*' Alternatively, the auditor could focus on the procedure's effectiveness by reporting: '*The corrective action procedures (Ref. XYZ) do not provide for the actions taken to be recorded and therefore verification of such actions cannot be verified in advance of subsequent incidents.*'

The example above concerning management reviews was not a nonconformity but it is worthy of being reported as an observation since ISO 9004-1 suggests the records contain conclusions and recommendations. Assuming that further checks revealed that the management review policy required records but did not define the content the observation statement would be worded as follows:

'*The management review records for June 93 and January 95 do not document the observations, conclusions and recommendations reached as a result of the review to indicate the suitability and effectiveness of the quality system as recommended by ISO 9004-1 paragraph 5.5, neither is there a policy requirement for such action.*'

Potential nonconformities may arise where instructions not yet implemented may cause nonconformity if implemented. An example is where an audit of an area planned in the future is also planned to be undertaken by someone who is not independent of the activities concerned, thereby causing potential nonconformity with clause 4.17 of ISO 9001. As the audit has yet to be conducted there is no nonconformity. The standard only requires that audits are carried out by independent personnel, not planned to be carried out by independent personnel.

Prevailing occupational conditions may impact on quality when hazards in the workplace will impair performance by injury, stress or health. These should be reported as observations.

Nonconformities require corrective action, whereas observations do not, so why report observations? Simply because observations do indicate that the effectiveness of the system is impaired and whilst no immediate action is essential, in time performance may deteriorate to a level below conformity. In most cases companies will act on the observations as well as the nonconformities since their aim is to improve their system. In fact observations are often welcomed and taken seriously.

Classification of nonconformities

Nonconformities are classified in order to prioritize corrective action and determine the nature of follow-up action. Classification is not essential. Some certification bodies in fact do not classify nonconformities and define nonconformities as hold points, a hold point being a situation which needs to be resolved before a certificate can be issued. Whether or not nonconformities are classified, a certificate cannot be issued until they are resolved. However, some certification bodies will issue a certificate of registration on being presented with an acceptable corrective action programme unless there are major nonconformities.

The classification of minor nonconformity has little emotive significance. However, a declaration that there are major nonconformities raises emotions, so the decision to classify any nonconformity as major should be taken with extreme care and should not be announced flippantly. A practical definition for the two terms is shown in the box opposite. Some other definitions of major and minor nonconformity are given in Boxes A & B[1]. These were obtained as a result of a series of letters to *Quality World* during 1994. Box C contains the only published definitions known to the author.

> **NONCONFORMITY CLASSIFICATIONS**
>
> *MAJOR*
>
> The absence or total breakdown of the provisions to cause product conformity or prevent product nonconformity with the expectations and needs of customers.
>
> *MINOR*
>
> Any failure to meet one or more requirements of the standard.

These definitions bear a strong similarity with those from BSIQA. The one problem with these QS 9000 definitions is that they do not define what a requirement of QS 9000 is. It could be an element such as 4.3, or a clause such as 4.3.1, a 'shall' statement or one requirement within a 'shall' statement. Having examined the companion *Quality System Assessment Guide*, it seems that a requirement is a 'shall' statement but it is unclear and some of the requirements include many 'shall' statements. It is also unclear what 'one item of a company's quality system' is. One item could be a trivial matter or an activity vital to achieving customer satisfaction. The same is true for the BSIQA definitions[2]. BSI define a clause as a section of the standard such as 4.1, 4.2, 4.3 etc. and a requirement as a sub-clause e.g. 4.6.2. In the IRCA exam paper for Lead Auditor Training courses, a clause is defined as a paragraph or section of the standard identified by a number such as 4.1.1.

One of the problems with basing the definitions on numbers of clauses and requirements of the standard is that the standard was not structured to facilitate this practice.

---

[1] *Quality World*, May 1995

[2] BSI are currently reviewing their definitions to relate them more closely to the effect on product or service.

A ALTERNATIVE DEFINITIONS OF A MAJOR NONCONFORMITY

1 A failure of the quality system which would adversely affect product quality

2 A demonstrated total absence or inadequacy of a necessary control arrangement throughout the supplier's organization

3 A number of failures of a particular control arrangement in different areas of activity with the supplier's organization demonstrating a clear failure of the system

4 The absence or the total breakdown of a system to meet the requirements of a clause of ISO 9000 or other referenced documents[3]

5 A failure to meet a contract requirement, department procedure or quality system requirement or a significant number of minor nonconformities which taken together constitute a threat to the system

6 A situation which undermines the purchaser's confidence in the quality of goods or services provided by the supplier

---

B ALTERNATIVE DEFINITIONS OF A MINOR NONCONFORMITY

1 A demonstrated absence of a control arrangement in one area of activity of the supplier

2 The failure of a particular control in one area of activity of the supplier which constitutes a quality risk

3 A number of noncompliances which when considered in total constitute a quality risk

4 An isolated failure to comply with a specified requirement

5 A situation which does not bear immediately on the purchaser's confidence in the quality of goods or services provided by the supplier

6 Either a failure to meet one requirement of a clause of ISO 9000 or other reference documents or a single observed lapse in following one item of a company procedure[4]

---

[3] As currently used by BSI Quality Assurance.

[4] As currently used by BSI Quality Assurance.

C   DEFINITIONS USED IN CONJUNCTION WITH QS 9000[5]

Major Nonconformance

- The absence or total breakdown of a system to meet a requirement of QS 9000

- Any noncompliance that would result in the probable shipment of a nonconforming product

- A condition that may result in the failure or materially reduce the usability of the products or services for their intended purpose

- A noncompliance that judgement and experience indicate is likely either to result in the failure of the quality system or to materially reduce its ability to assure controlled processes and products

Minor Nonconformance

- A noncompliance that judgement and experience indicate is not likely to result in the failure of the quality system or reduce its ability to assure controlled processes or products

- A failure in some part of the supplier's documented quality system relative to a QS 9000 requirement

- A single observed lapse in following one item of a company's quality system

---

A good example of this is in the 1987 version of ISO 9001 where the requirement for management review records was contained in clause 4.1.3 and the requirement for contract review records contained in element 4.3 as there were no subdivisions. In the 1994 version, the requirement for management review records remained in clause 4.1.3 but the requirement for contract review records was placed in a clause all by itself. If we apply our clause-based definitions of major and minor nonconformities to ISO 9001 1987, we see that an absence of management review and contract review records would be a minor nonconformity. However, in the 1994 version, the absence of contract review records now becomes a major nonconformity as it is the only requirement of clause 4.3.4. Ironically, nothing has changed except the structure of the standard.

Until we have an internationally agreed definition, the definitions I provide above are as good as any others. The important thing to remember is that whatever the definition, they should be agreed with the company at the outset and used by all auditors working for the particular second or third party.

---

[5] *Quality System Assessment* published by Chrysler, Ford and General Motors 1994

Determining system effectiveness

The effectiveness of a quality system is a measure of how well the system accomplishes its objectives. ISO 9001 states that the purpose of the system is to ensure that product conforms to specified requirements. This is one type of objective. With this simple objective the effectiveness of the system could be measured by how well it causes conforming product to be supplied and prevent nonconforming product from being supplied. However, a quality system may be designed to accomplish this and other objectives.

The defined quality policy should declare an intent to supply conforming product and the quality objectives may quantify a level of conformity to be achieved over a prescribed period, probably increasing year-on-year. The policy and objectives may also declare intentions regarding internal efficiency and effectiveness, the environment, supplier relationships etc. The effectiveness of the quality system could therefore be a measure of how well it causes achievement of the defined policies and objectives and prevents failure to meet such policies and achieve such objectives.

The word 'objectives' only appears twice in ISO 9001. Firstly in clause 4.1.1 on quality policy and secondly in clause 4.1.3 on management review. There is no specific requirement for the quality system to accomplish stated policies and objectives. This is in fact a recommendation of ISO 9004-1 and in view of the intent of ISO 9000, it would seem appropriate that any measure of effectiveness was against the organization's policy and objectives. Taking this approach has certain implications, however. Nonconformities with the standard may not indicate the system is ineffective; it depends on what they are. Also the system may not be effective even if no nonconformities with the standard are detected. It depends on what the policies and objectives are. A view often taken about ISO 9000 is that it permits companies to do what they want, providing it is documented. This claim can be justified when reading clause 4.1.1 of ISO 9001 where it merely requires the company to define and document its quality policy without requiring what the policy should address. So if a company claims that it is committed to quality but does not relate this to satisfying customers then the quality system can be effective with a high level of customer complaints. It is not until one searches through ISO 8402 that one realizes that quality means satisfying customer needs.

Internal efficiency and effectiveness may well be measured on a second party audit when the customer requires continual improvement in both quality and costs but is not normally measured on ISO 9000 certification audits. Determining system effectiveness therefore depends on the purpose of the audit:

- If there is an omission in a procedure such that certain activities or decisions required by the standard will not be carried out by implementing the procedure, then the system is ineffective as it cannot be relied upon to cause conformity.

- If staff use informal practices to cause conformity rather than the documented practices, then the system cannot be relied upon to prevent nonconformity or cause conformity.

- If there is a high probability that a nonconformity with a requirement of the standard would occur again, then the system is ineffective and corrective action is required.

- If the nonconformities are random and infrequent and caused by human error that additional controls, new technology and training would not remedy, then there is no fault with the system. Even in a 'zero defect' environment it is probably untrue to claim that no errors are ever made. To err is human after all.

- If objectives are not being achieved due to inadequate planning and implementation, then the system is ineffective.

It follows therefore that to determine whether or not the quality system is effective, regardless of the number of nonconformities, an analysis of both the nonconformities and observations needs to be performed. The conclusion as to whether the system is effective should be contained in the summary of audit findings (see below).

Drafting the audit report

There are four phases to the completion of the audit report:

1 A draft report indicating the results of the audit submitted to the company immediately following the audit.

2 An initial approved report as above but with the signature of the Management Representative and a commitment to submit corrective action proposals by an agreed date issued on completion of the Closing Meeting.

3 An intermediate report as above but with the addition of corrective action proposals from the company which have been evaluated by the auditor and issued on their acceptance.

4 A final report as above but with the addition of a declaration by the auditor that the corrective actions have been effectively implemented and issued on completion of the audit.

The Lead Auditor should co-ordinate the preparation of the draft report ensuring that it is both accurate and complete before being presented to the company. Following the Closing Meeting, the Lead Auditor should continue to co-ordinate inputs from the company and the auditors evaluating corrective actions.

Report content

The initial audit report should contain the results of the audit and as such should relate to the purpose and scope of the audit defined at the Opening Meeting. For example, if the purpose and scope of the audit was to establish the extent to which the quality system of Byte Computing Inc. of Santa Clara meets the requirements of ISO 9001 for the design, development, production, installation and servicing of mainframe computers and associated peripheral devices, then evidence should be presented in the report which demonstrates whether or not this objective has been achieved. The audit report should therefore include:

- A conclusion – included in the summary

- The evidence supporting the conclusion – contained in audit finding sheets

- The success criteria on which the conclusions are based – included in the summary

- The consequences of the audit findings – included in the summary

- The agreed follow-up action – included in the summary

- The audit plan as an appendix

- The site plan as an appendix

- Disclaimer statement – included in the summary

- Confidentiality statement – included on the front sheet

- Identification data – included on the front sheet

- Approval of report – included on the front sheet

This suggested contents for the audit report go beyond that recommended in ISO 10011. ISO 10011 describes the audit report as that which is submitted at the end of examining the organization's operations. Certain other agreements have been or will be made before the audit team leaves the site, either at the Opening Meeting or the Closing Meeting. Such agreements are:

- *The success criteria.* Having stated this at the Opening Meeting it should be recorded for future reference. Many certification bodies do not include this, perhaps for fear that they may be challenged at a later date. Its inclusion doesn't mean that it cannot be changed on subsequent audits, only that it is a point of reference.

- *The date by which corrective action proposals are to be submitted.* This will be agreed at the Closing Meeting and so should be recorded in the approved version of the report before leaving the premises.

- *The agreed follow-up action*. The precise follow-up action will depend on the outcome of the audit. Whatever the follow-up action, it should be recorded to avoid misunderstanding.

If these additional points are not included in the audit report, some minutes of the Closing Meeting need to be taken as evidence of the agreements.

Audit reports should not contain the names of personnel or confidential information if it can be avoided. Neither should they contain dialogue or stories such as:

> '*In talking to the workshop supervisor, Alan Jones, I detected that his procedures were out of date. He said they were the latest issue but on checking the master list I proved to him that this was not so. This is in violation of ISO 9001 clause 4.5.2 and is therefore a minor nonconformity.*'

This statement contains several bad traits. It names the supervisor, it reports a conversation, it uses emotive language (i.e. 'in violation of') and shows the auditor to be arrogant by his saying that he proved the supervisor to be wrong.

Report format

The format of the audit report needs to be of a standard which presents the results of the audit in a clear and unambiguous form. It is a means of communication and by using a standard format, it will cause the auditor to disclose the minimum amount of information and thereby enable continual improvement for the particular auditing organization. The information identified above can therefore be presented in the following format:

- Front sheet containing the identification data, report approvals and confidentiality statements; see Figure 5-4

- Summary sheet containing the success criteria, good points, conclusions, recommendations, agreed submission date for corrective action proposals and the agreed follow-up action; see Figure 5-5

- Compliance matrix showing the extent of compliance with requirements; see Figure 5-6

- Audit findings sheet containing nonconformity and observation statements; see Figure 5-7

- The audit plan showing who audited what and when; see Table 2-1

- The site plan showing where the areas audited were located

	QUALITY SYSTEM AUDIT REPORT	Report Ref: Page of
Company name	**Address**	

Location(s) of audit

Date of audit:	Duration:

Purpose and scope:

Baseline Documents	
Quality system standard:	
Quality system documents:	
Other documents:	

Audit team	
Name	Name

Company representatives	
Name & Position	Name & Position

Confidentiality

Report approval		**Report Distribution**	
Lead Auditor:	Management representative	Original	**Company**
		Copies	Audit Team
	Signature		
Name:	Name:		
Date:	Date:		

Figure 5-3 Audit report front sheet

	AUDIT SUMMARY	Report Ref:
		Page of

Success criteria:

Good points:

Summary of nonconformities and observations:

Conclusions:

Recommendations:

Corrective action proposal submission date:

Follow-up plan:

Figure 5-4 Audit summary sheet

COMPLIANCE MATRIX

Clause	Topic	Document Audit	Implementation Audit									
			Business Management	Quality system management	Marketing	Order Processing	Order planning	Product/service generation	Procurement	Support services	Order fulfilment
4.1.1	Quality policy										
4.1.2	Organisation										
4.1.2.1	Responsibility and authority										
4.1.2.2	Resources										
4.1.2.3	Management representative										
4.1.3	Management review										
4.2.1	Quality system										
4.2.2	Quality system procedures										
4.2.3	Quality planning										
4.3.1	Contract review procedures										
4.3.2	Contract review										
4.3.3	Amendment to contract										
4.3.4	Contract review records										
4.4.1	Design control procedures										
4.4.2	Design and development planning										
4.4.3	Organisation and technical interfaces										
4.4.4	Design input										
4.4.5	Design review										
4.4.6	Design output										
4.4.7	Design verification										
4.4.8	Design validation										
4.4.9	Design changes										
4.5.1	Document and data control procedures										
4.5.2	Document approval and issue										
4.5.3	Document changes										
4.6.1	Purchasing procedures										
4.6.2	Evaluation of subcontractors										
4.6.3	Purchasing data										
4.6.4.1	Supplier verification at subcontractors										

Figure 5-5 Compliance matrix (sheet 1)

COMPLIANCE MATRIX

Clause	Topic	Document Audit	Implementation Audit									
			Business Management	Quality system management	Marketing	Order Processing	Order planning	Product/service generation	Procurement	Support services	Order fulfilment
4.6.4.2	Customer verification of subcontracted product										
4.7	Control of customer supplied product										
4.8	Product identification and traceability										
4.9	Process control										
4.10.1	Inspection and testing procedures & records										
4.10.2	Receiving inspection and testing										
4.10.3	In-process inspection and testing										
4.10.4	Final inspection and testing										
4.10.5	Inspection and test records										
4.11.1	Control of inspection, measuring and test equipment										
4.11.2	Control procedure										
4.12	Inspection and test status										
4.13.1	Nonconforming product control procedures										
4.13.2	Nonconforming product review and disposition										
4.14.1	Corrective and preventive action										
4.14.2	Corrective action										
4.14.3	Preventive action										
4.15.1	Handling, storage, packaging, preservation and delivery										
4.15.2	Handling										
4.15.3	Storage										
4.15.4	Packaging										
4.15.5	Preservation										
4.15.6	Delivery										
4.16	Control of quality records										
4.17	Internal quality audits										
4.18	Training										
4.19	Servicing										
4.20.1	Identifying need for statistical techniques										
4.20.2	Statistical techniques application procedures										

Figure 5-6 Compliance matrix (sheet 2)

![magnifying glass]	**QUALITY SYSTEM AUDIT REPORT**	Report Ref:
		Page of

Ref	Location	Guide

Findings (include subject, incident, requirement, requirement reference and precise location if necessary)

Classification	Major nonconformity ☐	Minor nonconformity ☐	Observation ☐

Acceptance

Auditor Date:	Management representative Date:
Signature	Signature
Name	Name

Proposed action (including the specific remedy, the action on others in the population and action to prevent recurrence with implementation dates)

Supplementary sheets ..

Follow-up action

Proposed action accepted	Date:	By:	Signature:
Proposed action verified	Date:	By:	Signature:

Form RF/QA/004 Issue 1. 10 July 1995

Figure 5-7 Audit findings sheet

The conclusions

Any investigation should result in the collection of sufficient evidence on which to draw conclusions. If the evidence collected is inconclusive then the audit has not been a success. An inconclusive result may occur if the audit had to be terminated due to uncontrollable circumstances. However, if the audit is terminated by mutual agreement due to major nonconformities being revealed before completion of the audit plan then a conclusion can be declared. Possible conclusions are:

- *Sufficient objective evidence has been found to demonstrate that the quality system meets all the requirements of ISO 9001 for the design, development and manufacture of ...*

- *Sufficient objective evidence has been found to demonstrate that the quality system meets most of the requirements of ISO 9001 for the design, development and manufacture of ... However, as the nonconformities discovered are all of a minor nature, the effectiveness of the quality system is not significantly impaired.*

- *Insufficient objective evidence has been found to demonstrate that the quality system meets the requirements of ISO 9001 for the design, development and manufacture of ... As some of the nonconformities discovered are of a major nature, the quality system has therefore been proven to be ineffective in enabling the company to achieve the defined objectives.*

Evidence supporting the conclusions

The evidence to be contained in the report to support the conclusions should include:

- A matrix showing which requirements were checked in which areas and found conforming and which were found nonconforming. Such a matrix will denote the extent to which the system was found compliant with the specified requirements. If the audit has been carried out against contract documents (as in a second party audit), a separate matrix for each document may be necessary. An example using ISO 9001 is given in Figure 5.6.

- A list of the areas or activities which were found to be particularly good in ensuring the supply of conforming product. One reason for including this is so that the audit is seen as being positive. Perhaps a more practical reason is so that subsequent auditors can verify whether the good areas have been maintained. A deterioration in such areas may indicate a change of management, a declining commitment or other weakness in the effectiveness of the system. These can be added to the audit summary or included as a separate sheet if extensive.

- The nonconformity statements indicating their classification and the agreement of the Management Representative. Nonconformity statements are essential to obtain agreement to corrective action. They may make provision for the company to specify corrective action in each case but it is not essential that they do this (see Figure 5-7).

- The observations indicating areas for improvement, potential nonconformities or hazards. Observations do not require action. However, they should be included so that the company may if so desired take some action which can be verified on subsequent audits. Some of the observations may indicate a potential nonconformity which needs to be checked for deterioration (see Figure 5-7).

The consequences of the findings

The consequences of the findings should be presented in terms of the recommendations that will be made either to the certification body if third party or to the customer if second party. As a result of the audit findings there will be three possible outcomes:

- Unconditional recommendation for certification due to no systematic failure of the quality system being detected.

- Conditional recommendation for certification due to one or more minor nonconformities being detected.

- Refusal to grant certification due to one or more major nonconformities being detected.

The recommendations should be stated on the summary sheet; see Figure 5-5.

Certification bodies differ in their criteria for certification. All seem to agree that the presence of a major nonconformity warrants a refusal of certification. Some will not award certification until all minor and major nonconformities are closed. Others will award a certificate on receipt of a commitment by the company to close the minor nonconformities, and a few will issue a certificate with minor nonconformities remaining open.

The audit plan

The audit plan should be included in an appendix to the report as evidence of the areas covered and the auditors concerned. If the plan had to be changed at the Opening Meeting or had been changed subsequently, then the revised plan rather than the original plan should be included. The inclusion of the audit plan is necessary so as to indicate to those responsible for any subsequent audits what was covered and who was involved. It enables these auditors to build upon previous success or modify the plan accordingly, thereby saving time in planning future audits.

The site plan

A copy of the site plan should be included as an appendix to the report, if initially provided. By marking the areas audited it will indicate the coverage of the audit and act as a signal to subsequent auditors, what to check next time and what has changed since the last visit. The addition of offices, workshops etc. may only become apparent when a current site plan is compared with a previous plan.

Disclaimer statement

This can be a standard statement concerning the limitations of the audit. The limitations are declared through a disclaimer statement such as:

'The audit was based on a limited sample of operations and although conformance with all relevant requirements of the standard has been tested, other nonconformities to those reported may exist.'

Confidentiality statement

This can be a standard statement concerning the treatment of information contained in the report. Confidentiality of the report is declared through a statement such as:

'Any information obtained during the audit will remain confidential and will not be disclosed to a third party without the prior agreement of the company.'

If a Non-disclosure Agreement has been signed by the auditors on request of the company, then this should be included in the report and a reference made to it in place of or in addition to the standard statement.

Identification data

The draft audit report should have a front sheet (see Figure 5-4) which identifies:

- The report reference number

- The company name and location

- The date and duration of the audit

- The purpose and scope of the audit

- The version of the quality system standard and company quality system documentation against which the audit was conducted

- The name of the Lead Auditor and members of the audit team

- The names of those company representatives attending the Opening Meeting and Closing Meeting and their positions in the company

- The distribution for the report

Preliminary acceptance of the audit report

Prior to reporting the results to the company formally at a Closing Meeting, it is prudent for the Lead Auditor to gain acceptance of the Management Representative to the findings. Individual findings may have been accepted at each of the company feedback meetings but it is not until the report has been drafted that the overall conclusions and recommendations can be announced. The Lead Auditor could dispense with a preview if the results do not hold any surprises and he/she is confident that the nonconformity statements will be agreed. However, it is often wise to run through the detail beforehand as being challenged by the Management Representative at the Closing Meeting can be a traumatic experience. If the Management Representative has agreed the findings beforehand it is likely that he/she will support the findings if other members of staff seek to challenge them later. However, it is not unknown for a change of heart to take place when faced with a rebellious upper management team.

At this preview, the guides ought to be present to confirm the findings although they may not know whether requirements of the standard have been contravened. This is the Management Representative's role. His/her knowledge of the requirements is likely to be greater than that of anyone else in the company and so his/her acceptance is essential. The Audit Finding sheets (Figure 5-7) should be presented to the Management Representative for acceptance and signature and any dispute dealt with before the Closing Meeting. If there is an impasse then the Lead Auditor has to decide whether or not to withdraw the particular findings. Withdrawal should only be considered if additional objective evidence is presented to cause the findings to be unsafe. Often the dispute is over words rather than facts and rewording the statement may well result in acceptance. If the dispute is over the classification then it could indicate a lack of understanding, or poor communication at the Opening Meeting. If the impasse continues, then the Lead Auditor can choose to go ahead with the Closing Meeting and do battle with management! If withdrawing a finding will not change the conclusions then, unless there is evidence to the contrary, reclassify the finding as an observation. Remember you may be outnumbered at the Closing Meeting. If you have done your job professionally there should be no problem.

The Closing Meeting

Purpose

The purpose of the Closing Meeting is to present the results of the audit, declare the conclusions and recommendations and obtain agreement on any further action required.

Attendance at the Closing Meeting

The Closing Meeting should be attended by all the auditors together with representatives of the company. The company representatives should include as a minimum:

- The Chief Executive or Managing Director to demonstrate commitment and agreement to the findings

- The Management Representative to support the findings and agree to corrective actions and the follow-up plan

- The managers of each area audited to agree the findings and indicate commitment to corrective actions

- The guides to support the findings

Closing Meeting agenda

The agenda for the Closing Meeting is given in the box opposite.

The sequence is important as the wrong sequence can lead to the company drawing the wrong conclusions. This will create problems in gaining approval to the report and will in all probability extend the meeting unnecessarily. There is a view that the detailed findings should be reported before the conclusions and indeed if the findings

---

CLOSING MEETING AGENDA

- Introduction of those not present at the Opening Meeting
- Expression of thanks to the company
- Confirm purpose and scope
- Good points
- Summary of findings including disclaimer statement
- Conclusions
- Recommendations
- Detailed findings (optional)
- Questions on the audit findings
- Corrective action proposals
- Follow-up action
- Confirm confidentiality
- Approval of audit report
- Endorsement of the manual
- Appeals

are few then the managers will not have long to wait. If the findings are extensive, most managers will be impatient for the result and so may interject with queries that cause the team to deviate from their plan. The key is to maintain control. By putting yourself in the auditee's shoes, you can anticipate the problems and prepare the agenda to overcome them.

Preparing for the meeting

After compiling the report certain checks should be carried out and an agenda prepared:

- Number the pages of the report.

- Check that all audit finding sheets have been signed by the auditor and accepted by the Management Representative.

- Check that all audit findings being reported are included.

- Agree on those findings that will not be disclosed.

- Check that there was a witness to each of the findings and if not how significant this may be.

- Check that the statements can be read out as written.

- Agree on whether the findings will be read out as written or summarized.

- If findings are to be summarized, how will this be done (you don't want one auditor summarizing and another going into detail).

- Agree on the order in which the findings will be delivered and who will deliver what.

- Identify those findings which are likely to be challenged and decide how challenges will be dealt with.

- Prepare an agenda in the sequence in which it will be delivered.

- Prepare or select written statements to read out against each item of the agenda.

Introductions

The Lead Auditor should greet the people gathered and thank them for attending the meeting. Anyone present who did not attend the opening meeting should be introduced. An attendance list should be circulated if necessary. Sometimes the CEO will only attend the Closing Meeting. Also managers who were involved in the audit may wish to attend to hear the results. It is important for the Lead Auditor to be seen to be in control of the meeting so if the environment is unsuitable, i.e. too small a room, too hot or too noisy, then before proceeding further a more suitable location should be sought.

It is not normal for an agenda to be circulated as it will have been prepared only minutes beforehand. However, so that everyone knows what will transpire, the Lead Auditor should explain the purpose of the meeting and outline the agenda. A typical opening statement may take the following form:

> *'Thank you for attending this Closing Meeting which has been convened to present to you the results of the audit, to give you the opportunity to discuss the findings and agree on any future action required.'*

It is then prudent to outline the agenda as follows:

> *'As Lead Auditor I will summarize the findings, give you our conclusions and recommendations and then my team will report the results from the areas they audited. May I therefore request that all questions be held until we have delivered the report. Following your questions I will request proposals for corrective action and then inform you what follow-up action we intend to take'*

Whilst the Lead Auditor is talking one of the audit team could be noting the names of those present or circulating an attendance list. The names need to be included in the audit report prior to the end of the meeting. It is prudent to request that questions be held until the report is delivered otherwise the meeting runs like a traffic light. Stop - Go - Stop - Go.

Thanking the company

The Lead Auditor should thank the company for their hospitality, the provision of guides and the transport, lunch and anything else they have provided. This item on the agenda is not the place to comment on the quality system.

Confirming the purpose and scope

The Lead Auditor should read out the purpose and scope as written in the audit report. The statement should be the same as made at the Opening Meeting. Any differences, especially omissions such as the particular products or services included or excluded may give rise to challenges on specific findings. It is not unknown for a CEO to claim that he didn't realize the audit did not include certain areas or products.

The good points

It creates the right kind of perceptions if the positive aspects are dealt with before the negative aspects. No matter how an auditor may phrase a nonconformity report, it is still perceived as a negative result. Emphasizing the good points will show that the audit team was not blind to all the hard work put in and it tends to put the other findings in perspective. If there are many good points then the Lead Auditor can summarize them and leave the detail to his/her team to amplify. However, do not be patronizing. Be genuine in your praise and speak with conviction. Most of all do not use comparisons with other companies that you might have visited. Even saying that 'you are better than most' may get back to your other companies and you may have some explaining to do.

Summary of findings

The Lead Auditor should summarize the findings by stating what was covered and indicating the number of minor and major nonconformities found rather than summarizing the content of the individual reports. A typical statement might be:

> 'We covered all areas identified in our audit plan which included the marketing, design, purchasing, production, QA and servicing departments and found no major nonconformities, 34 minor nonconformities and several observations.'

To this might be added the disclaimer statement:

> 'The audit was based on a limited sample of operations and although conformance with all relevant requirements of the standard has been tested, other nonconformities to those reported may exist'.

It is probably more effective at the Closing Meeting to indicate the departments covered by the audit rather than the processes denoted in the audit plan and they will be far more familiar to those present. If there are too many departments to mention, then say that all departments were included and name the exceptions. If there were several sites then name these.

Before going into detail on the findings it is prudent to declare the conclusions and recommendations, more or less following on from the summary statement.

Conclusions

The Lead Auditor should follow the summary of findings with a statement of conclusions. Typical statements were given earlier but for completeness one of these is repeated as follows:

> 'Sufficient objective evidence has been found to demonstrate that the quality system meets most of the requirements of ISO 9001 for the design, development and manufac-

ture of ... However, as the nonconformities discovered are all of a minor nature, the effec-
tiveness of the quality system is not judged to be significantly impaired.'

By following the summary of findings with the conclusions it prevents argument breaking out if the conclusions are favourable. If conclusions are not declared beforehand, each nonconformity may be challenged as the company may believe any nonconformity to warrant a refusal to certify. If the conclusions are unfavourable, then it may be a tough ride anyway. Most managers want summaries not details, so it is unreasonable to expect them to be patient whilst your team reads out 34 minor nonconformities unless you have forewarned them.

Recommendations

Following straight on from the conclusions the Lead Auditor should declare the recommendation that will be made to the certifying body (if third party) or customer (if second party). The recommendation is a consequence of the conclusion. It is not nor should it be a recommendation for corrective action. The three possible outcomes were dealt with earlier. Typical statements might be as follows:

'We are therefore delighted to be able to recommend the company for certification.'

or:

'We are therefore able to recommend the company for certification, conditional on the minor nonconformities being corrected to our satisfaction.'

or:

'We are therefore unable at this time to recommend the company for certification.'

Detailed findings

Before inviting questions the Lead Auditor should then invite each member of the team to report their findings. If there are as many as 34 minor nonconformities and indeed just as many observations, it may be prudent for the team to summarize since it could take hours to deliver the report (more like a budget speech in parliament!) Before the auditors deliver their reports, it is prudent to repeat a request that questions be held until the team have completed their report. The Lead Auditor can also suggest that note taking is welcomed so as to facilitate the questioning later.

If the major nonconformities are scattered amongst the team then the Lead Auditor should report these first. Also the Lead Auditor will have recorded findings as well but it is sensible to leave these until after the other team members have reported so that the ball is back in the leader's court before inviting questions. The sequence is important so as to convey the significance of the findings. If several auditors followed separate trails and found related evidence which when combined led to a classification of major, it again may be prudent for the Lead Auditor to deliver these findings in the interests of continuity. Whatever the game plan, the leader should work this out beforehand so that the delivery runs smoothly.

It cannot be stressed too strongly that the audit findings should be read from the report. It is dangerous to ad lib, amplify, explain etc. All this does is provide opportunities for the auditee to question the findings and cast doubt on the conclusion. In preparation for the Closing Meeting each auditor should read aloud the statements they have written to check the smoothness of delivery. Details such as serial numbers, procedures references, revision status etc. can be omitted in the verbal delivery if it aids a smooth delivery. But be careful not to omit material facts. This is one reason for including the clause reference in the statement since to place it elsewhere on the form may cause an interruption in the flow, or as so often happens, the requirement not met is not even mentioned.

> **AUDITOR'S VERBAL REPORT**
>
> - What areas were audited
> - Any good points worthy of note
> - Number of nonconformities and observations found.
> - The major nonconformities
> - The minor nonconformities
> - The observations
>
> **AND**
>
> **Never report audit findings that have been previously withdrawn by the Lead Auditor**

If summarizing the audit findings rather than reading them out one by one, the auditor can adopt the following style:

- *No final inspection records for PCBs as required by clause 4.10.4.*

- *Audit not carried out by independent auditor as required by clause 4.17.*

- *Quality objectives not documented as required by clause 4.1.1.*

- *Inspectors on PCB line not properly trained as required by clause 4.18.*

Questions on the audit findings

Once the report has been delivered, questions can be invited from the company. It may be unreasonable to expect that those present will have remembered the detail especially if 34 nonconformity statements and just as many observations were read out. However, some will have taken notes. If the findings have been summarized in the style above, then when questioned, the auditor should read out the full statement and ask if it is clear.

If the auditor has followed the rules given earlier then there should not be questions such as:

- *Where does it say that in the standard?*
- *What does the standard require?*
- *Do we have to do that?*

If questions of validity arise this is where the guide can assist in confirming the facts. When challenged, the auditor can indicate which guide was present and leave it to the questioner to question the guide. If you didn't have a witness then hard luck. You should have known better and prepared for the meeting properly. A number of ways of dealing with challenges are given in Chapter 4.

If asked to explain the finding then don't assume this is a challenge. Also do not take the opportunity to tell a story of how it came about. It wastes time and you may disclose something which conflicts with your written statement. Ask the questioner to explain what it is about the nonconformity statement that is unclear. If a straight-forward response doesn't satisfy the questioner, turn the statement around another way. Break it up into small parts. Refer the questioner to the standard. Put it in context by stating the purpose of the requirement and the consequences of nonconformity making sure your explanation does not conflict with the standard. Above all do not add trivia or go on the defensive. Be firm, confident and polite.

Questions that may arise are requests for advice on how the nonconformity could be corrected. An auditor must not recommend a corrective action to overcome a particular nonconformity. This does not mean that suggestions for improvement cannot be given or a range of options presented. See Chapter 4 for a more detailed analysis.

Corrective action statements

Requesting corrective action

When questions have been dealt with, the Lead Auditor should request the company to make a proposal for corrective action. If the findings are few, the proposals may be dealt with on the spot and added to the relevant forms. If the company would like some time to think about the action it intends to take then a date by which the proposals will be submitted should be obtained. Typical statements might be:

> 'Action on the observations is not required for certification purposes but certification does depend upon action being taken on the nonconformities. I would now like to invite the company to propose the action they intend to take to resolve these nonconformities. Proposals may be submitted now or by a date upon which we can agree.'

The date by which corrective action proposals will be submitted to the auditor should be denoted in the audit report (see Figure 5-5).

A Chief Executive's Tale

Jun made his way to the Chief Executive Officer's office, led by his guide Jess. This was Jun's first experience of interviewing a CEO and he approached the task with some trepidation.

After warm greetings from the CEO, Jun started his questioning. He asked the CEO if he knew the company's quality policy. The CEO answered that he did and so Jun proceeded to ask where it was documented. The CEO answered that it was in the Quality Manual. Jun had the Quality Manual in front of him and read the policy statement. Satisfied that this indeed was the quality policy, Jun proceeded to ask about the CEO's responsibilities and again the CEO pointed Jun to the statement in the manual. After several further questions in the same vein, the CEO turned to Jun and asked him if he had read the Quality Manual. Jun answered that he had. 'In that case,' said the CEO, 'why are you asking me questions to which you know the answers and particularly when the answers are contained in the document in front of you? Are you trying to catch me out?' Jun, taken aback by this response, apologized and proceeded to consult the standard.

Meanwhile Jess, who had witnessed many audits, whispered to Jun that it might be a good idea to ask the CEO about his customers and how he determined their expectations and the resources needed to meet them. Not wanting to appear a fool, Jun checked his standard first and found that such a question would in fact be a good one, but was puzzled why it was not on his check list.

Jun did ask about customer expectations and resources and found the CEO most co-operative. He also found that no resources had been allocated for maintaining the quality system following certification and so had revealed something of which the CEO was unaware.

The moral of this story is:

When talking with top management, talk about matters that interest them, otherwise don't be surprised if you are perceived as a fool!

Criteria for effective corrective action proposals

Auditors often forget that corrective action is action taken to eliminate the causes of an existing nonconformity to prevent its recurrence (Ref. ISO 8402 item 4.14). Action taken to eliminate the nonconformity is a remedial action (although ISO 8402 refers to this as correction) and many actual corrective action requests only require remedial action as there is no systematic fault. Internal auditors should also note that clause 4.17 of ISO 9001 calls for management to take timely corrective action and not simply to remedy the deficiency.

Fixing the piece of equipment without a calibration status label is only a third of the action required. The action required to address any nonconformity effectively has three components.

The Lead Auditor should explain that the action needed should:

- Correct the specific nonconforming item, i.e. the subject of the audit finding

- Seek out and correct any other similar instances of nonconformity

- Correct that which caused the nonconformity

There may be other items in a similar condition. The audit was only of a sample of operations so other nonconformities may well exist. The company should seek these out and correct them. This action, however, is not always appropriate, particularly when only one such item exists.

If the rules about reporting nonconformities have been followed, then only systemic failures will have been reported, i.e. failures for which corrective action is possible. There will therefore be a cause of the nonconformity and the company needs to establish what this is and specify how it can be eliminated. The why-why technique can be useful in revealing the root cause. Asking why seven times normally gets to the root cause. However, auditors should examine corrective proposals carefully to establish that the root cause has been identified and its elimination will in fact prevent recurrence of the nonconformity. Sometimes the root cause may be a lack of training. As all nonconformities are caused by people, we could argue that retraining would solve them all but retraining may not be the solution to preventing recurrence. Even with trained personnel, specifications, plans, procedures etc. are needed to ensure trained people do the right things. Trained people are not necessarily willing people. There may be something in the culture which is at the root of the company's problems.

Auditors should develop some idea of the root cause of the nonconformity during the audit. If it is not obvious, asking the auditee may reveal some useful suggestions.

Examples of corrective action proposals and how to evaluate them are given in Chapter 6.

The audit team may be presented with evidence of some action that has already been taken to correct the nonconformity. What should the auditor do? It very much depends on the nature of the original nonconformity:

- If the problem was a difference between practice and procedure and the practice was correct, then it is quite possible that the nonconformity could be closed.

- If the problem was a difference between practice and procedure and the procedure was correct then a further visit will be necessary to confirm the claims of the company.

- If the problem was an omission in the system such that neither procedure nor practice was compliant then a further visit will be necessary. Being offered the revised procedure is insufficient evidence that the system has changed. Records of effective operation of the system need to be examined.

Completion dates

The action to be taken is one thing. The date by which it is proposed to complete it is another. There may well be three different dates for each of the actions to be taken. Often all three are grouped together and one date proposed. It depends on the nature of the problem. If the nonconformity remains, how likely is it that product quality will be affected? If it is highly likely, or there is current evidence of this, then the remedy has to be immediate and the search for other nonconformities completed within a few days, a week at the most. Timing of the preventive action will depend on the company's current orders, contracts, design and production plans etc. There may be little prospect of it recurring in the next couple of months so the preventive action could be set at two months. This is where the auditor needs to judge significance and priorities. Rejecting a completion date of three months when the problem is minor may be inappropriate. Accepting a completion date of three months when the problem is major may be also inappropriate.

It has become custom and practice for up to three months to be given to resolve nonconformities. Any more and it indicates that the system is immature or some other deep-rooted problem. Completion dates should be denoted in the specific audit finding sheet (see Figure 5-7).

Follow-up action

The Lead Auditor should declare what follow-up action will be taken. This will depend upon the effect of recommendations:

- An unconditional recommendation requires no follow-up action.

- A conditional recommendation requires follow-up action which may vary from resolution through correspondence to closure of the nonconformity during the first surveillance visit.

- A refusal to grant certification requires a repeat visit to either examine that part of the system which is nonconforming or to repeat the complete assessment where there are several major nonconformities that warrant system re-design.

Follow-up action covers:

- The submission of the corrective action proposals

- The evaluation and acceptance of these proposals

- Verification that the nonconformity has been eliminated and prevented from recurrence

If the practices conformed to the standard but there were problems with the corresponding documentation, then it may be practical to verify implementation of the nonconformities through the postal service and for this to be confirmed at the first surveillance visit. If the nonconformities are minor, it may be practical to verify implementation of corrective actions at the first surveillance visit and initiate certification on receipt of satisfactory corrective action proposals.

If the nonconformities are major, then a repeat visit will be necessary to examine that area in which the major nonconformity was detected. Depending on the nature of the problem, the examination may be isolated to one particular area but may extend to other areas and involve a partial system audit. If resolving the problems necessitates a re-design of the system then the full system audit should be repeated. A repeat system audit may be outside the scope of the contract between the certification body and the company so the Lead Auditor may have to make the company aware of the cost impact.

The agreed action should be indicated in the audit report so that the company have a reminder of the auditor's intentions.

Confirming confidentiality

The Lead Auditor should repeat the confidentiality statement made at the Opening Meeting.

Distribution of audit report

The Lead Auditor should obtain agreement on the distribution of the audit report. It is normal practice to indicate that the original report will be left with the company with copies issued to the audit team only, thereby preserving confidentiality. However, any copies the company takes are its affair. Should the auditor's organization require additional copies then these should be agreed with the company at the Closing Meeting.

Approval of the audit report

When all other business has been transacted, the Lead Auditor should offer the audit report to the Management Representative for signature. If the meeting has gone well this should be routine. If there have been problems, there may be a reluctance to sign. This is where the Lead Auditor's selling skills are required. Whether or not a signature is obtained, the Lead Auditor could explain that the certificate will not be issued until the nonconformities are resolved to the satisfaction of the certification body or customer.

Endorsement of Quality Manual

It is common practice, but not essential, for a copy of the quality manual used for the assessment to be endorsed by the Lead Auditor by stamp or signature, thereby acting as a baseline against which subsequent changes may be verified. This manual should be retained by the company and be made available to the audit team on future audits.

Appeals

If a signature cannot be secured then the Lead Auditor should explain that there is an appeal procedure. The company can appeal against the decision of the Lead Auditor to the certification body or the customer if a second party audit. If the company gets no satisfaction from this avenue, an appeal to the accreditation board is the next stage.

Leaving the company

Before leaving the company ensure that any company documents are left behind together with the protective clothing etc. that has been provided. Request a copy of the audit report and express again your appreciation. Your closing remarks might be as indicated in the box.

> **CLOSING REMARKS**
>
> *We hope that you have been satisfied with the manner in which the audit has been conducted. If you do have any suggestions for improvement may we recommend you send them to our Cardiff office. We realize that you have a choice of certification bodies and thank you for choosing Assurance Technology International.*

Summary

This chapter has addressed the reporting process which commences once the audit findings have been recorded. The following are the key issues covered:

◻ In reporting the findings the key objective is to establish the effectiveness of the quality system, not to reveal everything that occurred during the audit.

◻ The best time to obtain agreement on nonconformities is at company feedback meetings and these should be held at the beginning of the next day of auditing.

◻ The difference between a major and a minor nonconformity is that a major nonconformity is a failure to cause product conformity or prevent product nonconformity.

◻ Corrective action is requested when failures can be prevented from recurrence, not when mistakes are found.

◻ To be valid, nonconformity statements should disclose the subject, the location, the incident and the requirement not met, avoiding emotive language.

◻ If there is insufficient evidence of nonconformity but system effectiveness is impaired then report as an observation.

◻ System effectiveness is a measure of how well the system accomplishes its stated objectives, not how many nonconformities were found.

◻ Audit reports should only contain the facts, not emotive language and conversations.

◻ The main reason for a Closing Meeting is to obtain commitment from the company to corrective action on the findings.

◻ The conclusions of the audit should relate to the effectiveness of the quality system and should be supported with objective evidence.

◻ It is better to read from the report than to ad lib and elaborate, as this will only cause doubt and conflict.

◻ Reporting good points and the degree of conformity provides a balanced report.

❏ Team members should report their findings, as it comes from the horse's mouth and avoids embarrassment.

❏ Corrective actions should prevent the recurrence of nonconformities, not just resolve the specific incident.

❏ Follow-up action depends upon the results of the audit and can range from no visits to a complete re-assessment.

❏ An expression of gratitude for the facilities and help received creates a cordial note before leaving the site.

Chapter 6

Quality system surveillance

The surveillance process

The surveillance process is similar to the process for conducting the initial audit. Quality system surveillance[1] has to be planned and the conduct and reporting process will be the same, except that depending on the type of surveillance, the Opening Meeting may be less formal. Many of the steps in the initial audit can be omitted, as information on the organization will already exist. A simplified process is illustrated in Figure 6-1.

Purpose

The purpose of conducting periodic surveillance is to establish that the quality system is being maintained and performing the function for which it was designed.

It is not unknown that following initial certification, companies turn their sights to other ventures and allow the quality system to decay. Some organizations only pursue certification for the certificate. They treat it as a goal and once won, they don't think they have to do anything to maintain the standard. As with any dynamic organism, the status quo never remains still for long. Even stability has to be maintained, otherwise forces outside one's control will destabilize the ship. The surveillance audit fulfils this function. It is perhaps the most important aspect of the quality management programme and if done properly can be of increasing value to the organization. If done

---

[1] The term 'quality surveillance' is defined in ISO 8402 and in some texts is referred to as 'continuing assessment'.

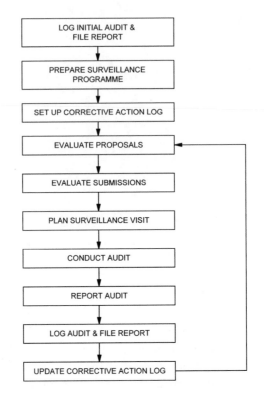

Figure 6-1 Surveillance process

poorly, it turns into a charade of little value. If there is one aspect which differentiates quality system certification from quality awards, it is the continual surveillance of the assessed quality system.

Surveillance consists of:

- Following up the actions agreed on the previous audit

- Verifying that standards have been maintained

- Verifying that the data generated by the system is being used for continual improvement

- Establishing if any changes need an extension to the scope of registration, the scope of the audit or a complete system audit

- Performing partial re-audits if a rolling audit programme is in force (see below)

Logging the system audit

On return from the initial audit or subsequent system audit, the audit report should be logged against the overall audit programme for the certification body. This will indicate one more objective achieved. The report should be vetted by the Lead Auditor's superiors and action taken to initiate registration or renewal of registration as appropriate. The practices of each certification body differ and as an administrative issue it is not of concern here to discuss these practices.

Preparing the surveillance programme

Surveillance strategy

The surveillance programme will depend on the size and complexity of the organization, the results of the initial audit and the surveillance strategy being adopted. It is usually a condition of registration that the complete quality system be audited once every three years and for monitoring of the system to be performed in the interval between assessments. This can be achieved in one of three ways:

- Periodic assessment, in which the quality system is subject to system maintenance monitoring between periods and complete re-assessment at the end of each period.

- Extended assessment, in which those parts of the organization, its products and processes excluded from the full assessment sample are subject to audit on each visit together with system maintenance monitoring and a complete re-assessment at the end of each period.

- Continual assessment, in which selected parts of the quality system are assessed on each visit and which over a given period subject the whole quality system to re-assessment. In addition, monitoring of improvement and maintenance elements of the system is carried out.

Comparing surveillance strategies

The three strategies are compared in Table 6-1. One industry sector may benefit more from one strategy than another. The best strategy may be a combination of all three.

Topic	Periodic assessment	Extended assessment	Continual assessment
Suitability	Unstable industries	Complex industries	Stable industries
Coverage	Requires several years to cover every operation in large companies	Covers all operations over shorter timescales	Covers all operations in less time than periodic option but more time than extended option
Client readiness	Areas may become complacent	Keeps most areas vigilant	Keeps some areas vigilant
Degree of disruption	Little between full assessments but major disruption every 3yrs	Continual disruption but on a reduced scale	Disruption in one area each visit
Use of audit resources	Disperses resource	Disperses resource	Concentrates resource in one area
Number of audit days	Minimum level between full assessments	Higher than minimum between full assessments	Constant level can be maintained
Tenure	Annual contract	Annual contract	3 year contract
Interfaces	Tests interfaces between processes	Tests interfaces between processes	May not test interfaces between processes
Test of capability	Periodic test of capability	Continual test of capability	May not test capability
Audit strategy	Can be process based	Can be process based	Tends to be department based

Table 6-1 Surveillance strategies compared

Frequency

The frequency of the surveillance visits will vary depending on a number of factors:

- The results of the initial audit
- The perceived commitment of the company
- Notification of changes in the quality system
- The timing of rolling re-audits if third party
- The criticality of supplies if second party
- Distance, size and complexity

> **SURVEILLANCE STRATEGY OPTIONS**
>
> - Periodic assessment:
>   3yr full assessment + monitoring
> - Extended assessment:
>   3yr full assessment + monitoring + extended audit coverage
> - Continual assessment:
>   Partial system audit + monitoring

It is normal practice to schedule two surveillance visits each year regardless of the organization. This has the advantage of enabling the auditor to build trust with the personalities in the organization. It is also likely that many of the people met on the initial visit will still be around. Push it out to once a year or further and faces change such that the auditor has to renew communication. However, it is prudent to plan more visits after the initial audit, then reduce the frequency as confidence increases. This way the people in the organization will feel that they are affecting the frequency of visits and hence reducing their costs.

As any audit only takes a sample of operations there will be some operations that have not been audited and with new insight into the way the company operates, the auditor can prepare a meaningful and comprehensive surveillance programme.

Even on a surveillance visit, areas not audited previously should be examined so that over the three-year period all areas are visited. This is somewhat different to the rolling re-audit where there is no periodic full system audit.

Selecting the auditor

It is prudent to select the same auditor or Lead Auditor for surveillance as for the initial audit but there are pros and cons to be considered depending on the duration and type of relationship that has been formed:

- It allows mutual trust to be established between the auditor and auditee.

- It also cuts out any learning necessary and so saves time.

- Clients prefer a face they know.

- Too cosy a relationship can lead to the auditor taking things for granted and making assumptions.

- Changing the auditor can be good for the client as it enables a new perspective to be applied.

- Changing the auditor can be good for the certification body as it facilitates the calibration of their auditors.

- For team audits, keeping the same lead and changing the team members can be beneficial to the leader.

- If there have been complaints then a different auditor should always be chosen.

Corrective actions

The corrective action log

The auditor or his/her organization should establish and maintain a corrective action log for each company. This will enable them to track outstanding corrective actions and prompt the company when completion dates are overdue. The log will also enable the auditor to warn the company of impending withdrawal of certification should they be repeatedly delinquent in responding to requests for corrective action. A simple log might contain the following:

Audit Finding Ref	Report date	CA proposal submittal date	Date proposal accepted	Corrective action completion date	Follow-up visit due date	Date corrective action verified
DH 001	10 July 95	31 July 95	2 Aug 95	1 Sept 95	5 Oct 95	5 Oct 95
DH 002	10 July 95	31 July 95	2 Aug 95	31 July 95	None	5 Oct 95

Establishing trends

Another form of log is a record that will enable trends to be established. The above log will aid this purpose in so far as timeliness but not with respect to the nature and frequency of the problems encountered. With a computer database many statistics can be compiled to establish:

- Whether the number of nonconformities is increasing or declining in particular areas

- Whether the incidence of nonconformity against particular requirements is increasing or declining

- Whether particular processes, people, products or procedures give rise to frequent problems

- Which areas are stable and which volatile

These and many others will help focus attention on the areas where improvement can be made and where it can be demonstrated that audits add value. If an auditor does not take these factors into account then it is likely that they will not focus on the productive areas and as a result acquire the reputation of being inflexible, old fashioned or just plain incompetent.

Evaluating corrective action proposals

Corrective action proposals may be offered at the Closing Meeting of the initial audit. However, it is more frequent for the company to want to think carefully about the results and take time to consider the most appropriate action. Whenever the proposals are offered, the auditor needs to evaluate them against predefined criteria. The criteria were defined in Chapter 5 and required the corrective action to:

- Correct the specific nonconforming item, i.e. the subject of the audit finding

- Seek out and correct any other similar instances of nonconformity

- Correct that which caused the nonconformity

When evaluating the proposals, take time to analyse their potential effect. Has the root cause been addressed? Are there likely to be other causes which if not addressed may cause the problem to recur? With most problems the key factors that influence their cause are likely to be the same as those used to establish that the organization has its operations under control. Namely:

- People. Do they need training, do they receive feedback. Even if the procedure is fixed, do the right people have the right authority to make it work?

- Product. Was it made incorrectly or was it designed incorrectly?

- Service. Are the requirements specified?

- Documents. Will changing the procedure create conflict with the policy? If one document changes are others affected?

- Data. What data is affected? If procedures change will the data be changed?

- Records. Does the format of the records need to change, do additional records need to be designed?

- Materials. Are the requirements specified? Will the supplier know?

- Measurement. Will the new provisions fall within the scope of existing verification procedures?

- Tools/Equipment. New procedures may need new tools and equipment – have they been specified?

- Resources. Does the new practice require additional or different people, equipment, facilities etc.? Will something else degrade as a result?

These are only a few of the factors the auditor needs to consider and quite often the proposals will not contain sufficient information for full consideration to be given. Correcting nonconformities usually requires a change to the system. Even enforcing compliance may need a wider distribution list for the information. A change in one area may cause problems in another, that is why it is insufficient just to look at the specific nonconformity.

One effective way of analysing the problem is to draw a cause and effect diagram using the nine key factors above. Often such elaboration is unnecessary but a perceptive auditor can raise his/her credibility enormously if in reply to the proposals, unidentified major contributing factors are revealed.

The auditor should respond to the company's proposals with encouragement even if they are inadequate. However, be careful not to tender advice with specific solutions. The proposals will either eliminate the problem, deal with the quick fix only or adversely affect another process.

Evaluating corrective action submissions

Some of the nonconformities may be dealt with by correspondence, particularly those requiring changes to documentation. If the practice was compliant but was not reflected in the documentation then confirmation that the documentation has been changed properly can be obtained off-site. Also, records that were unavailable at the time could be submitted later and confirmed adequate. Examples may be minutes of a management review meeting, a mislaid audit report or a mislaid design review report.

If a formal visit is being waived in view of the minor nature of the nonconformities, then closure of the nonconformities depends upon the results of such an evaluation. If a formal visit has been planned anyway to examine other corrective actions, then a final decision of the acceptability of the submission can be postponed until on site.

In examining such submissions in lieu of making a formal visit the auditor should verify:

- That the action taken matches that which was proposed

> **CORRECTIVE ACTIONS**
>
> *Corrective actions should:*
>
> - Correct the specific nonconformity.
> - Seek out and correct any other similar instances of nonconformity.
> - Correct that which caused the nonconformity.

- That the action taken eliminates the cause of the nonconformity

If the action taken is slightly different from that which was proposed, then assess the difference and, if questionable, query the change with the client. It may be that on closer examination the client found a better solution. It may also be that they found it too expensive or politically inexpedient. If the latter reason is suspected then check very carefully that the requirements of the standard have been met and that there is no impact on other areas which have not been addressed.

When examining records, they could be fabricated for the occasion, but it is unlikely. If the records seem far too contrived, then reserve judgement until the next site visit. Fabrication of evidence can stick out like a sore thumb when compared against similar records that existed at the time of the previous audit.

Planning the surveillance visit

The surveillance plan will depend upon the adopted surveillance strategy but should be planned in the same way as the initial audit, apart from those activities that do not need repeating.

A typical planning sequence is illustrated in Figure 6-2. The audit schedule will be similar to that illustrated in Chapter 3 except that for simple monitoring between complete assessments, there may be no formal schedule.

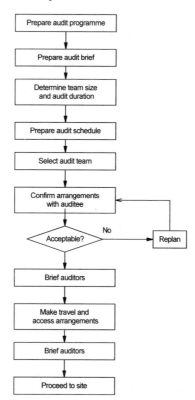

Figure 6-2 Surveillance planning process

The Bribe That Didn't Work

John, the Lead Auditor, met with the Directors to determine their commitment to quality and establish how they were using the quality system to ensure customer requirements were met. He entered the Managing Director's office with his guide Eugene the Quality Manager. Having been greeted by the MD, the Industrial Relations Manager and the Commercial Manager, John proceeded to ask questions about the organization's vision and values, hoping to lead on to quality matters once he had won their confidence.

To his surprise, the MD was very informed about ISO 9000 and how it could be used to help his business prosper. John noticed that the MD was doing all the talking so he tried to entice the other directors to join in the discussion. At this point John's pen started to write intermittently so Martin, the Commercial Manager, offered his pen to enable John to continue writing.

Thanking Martin for his kindness, John continued his questioning. After about an hour John was satisfied that he had obtained sufficient evidence of compliance and thanked the directors for their contribution.

Later, when moving between interviews, the Commercial Manager spotted John and called him over to hand him a parcel. Martin explained that as John's pen was obviously not of adequate quality he might like one of the company's presentation pens to help him through the audit. John, surprised at such kindness, accepted the pen gratefully and went on his way.

During the late afternoon, John started to prepare the nonconformity statements of the day. Eugene, eager to see the results of the audit so far, was surprised that John had found so many things to write about. In fact John had generated some 20 nonconformity reports.

When Eugene met Martin the following morning, he reprimanded him, claiming that the gift of a pen had only helped John to write more nonconformities.

At the Closing Meeting, after listening to John's report, Martin was surprised that John had found so many nonconformities in his area, particularly as they had been written with the pen he had given John. Eugene made light of this matter by saying that it was mere coincidence and that John was obviously not a person that could be bribed.

The moral of this story is:

A gift is a gift except when intended as a bribe!

Conducting surveillance

The surveillance visit should be conducted in the same way as a formal audit. However, if the adopted strategy is for periodic assessment, then surveillance visits will be limited to monitoring. When this is the case there is little need for a formal Opening and Closing Meeting with upper management present. The Opening and Closing Meetings can be limited to the Management Representative and the auditor but should never be dispensed with. Agreement to the plan and the findings is essential in all audits.

In covering each area the auditor should address:

- The previous nonconformities

- The improvement elements of the system

- The maintenance elements of the system

- The aspects to be audited / re-audited in the chosen area

Previous nonconformities

These can be dealt with collectively with the Management Representative or in each area as they arise. The auditor should request a report on the status of the corrective action and, if indicated complete, should check its effectiveness. Effectiveness doesn't mean that all that has to be done is to check that the specified corrective action has been carried out. The specified action may have been completely ineffective. Evidence should be gathered to show that no nonconformity with the governing standard exists.

Monitoring

In all three types of surveillance, monitoring is performed of the maintenance and improvement elements of the quality system. These are:

- Quality system documentation (clause 4.2.2), particularly changes since last audit

- Design review (clause 4.4.6), particularly feedback data from previous designs

- Corrective action system (clause 4.14.2), particularly customer complaints, product, process and system nonconformities

- Preventive action system (clause 4.14.3), particularly trend analysis of product, process and system performance

- Training programmes (clause 4.18), particularly their implementation and maintenance of training records

- Management review (clause 4.1.3), particularly the decisions made as a result of an analysis of system data

- Internal audit programme (clause 4.17), particularly its depth, coverage and frequency of audits

- Document control (clause 4.5), particularly document approval, change and version control

- Quality policy and objectives (clause 4.1.1), particularly their continued relevance to organizational goals and expectations of customers

- Process control (clause 4.9), particularly the monitoring of product and process characteristics and process capability studies

- Quality records (clause 4.16), particularly their indexing, storage and maintenance.

SURVEILLANCE CHECKS

• Quality system changes	• Internal audit programme
• Quality policy	• Quality objectives
• Preventive action system	• Corrective action system
• Management review	• Training programmes
• Document control	• Design reviews
• Process control	• Quality records

Some of the more common weaknesses found on surveillance visits are as follows:

- The internal audit programme has not been maintained.

- The frequency of audits has not been adjusted to take account of the status and importance of the audited activities.

- The management review frequency is fixed and has not taken account of changes in the technology, the organization or the market.

- Customer complaints are being dealt with but no corrective action taken to prevent recurrence.

- Analysis of the data generated by the system is not being used to predict problems and take preventive action.

- The training needs analysis has not been repeated and the training programme has been cut back.

- Procedures have not kept pace with changes to operating practices.

There is a suggestion in some quarters that surveillance could be limited to this type of monitoring and that the auditor could rely on the results of internal audits to obtain confidence in the system's effectiveness. Certainly, the auditor will examine the internal audit records and use the evidence to support conformity in areas where specific operations cannot be witnessed at the time of the audit. However, to place total confidence in the internal audit programme requires an in-depth knowledge of the culture, management style, long-term performance and the competence of the personnel involved. If indeed such confidence can be obtained, then there is no reason why surveillance should not be reduced. Confidence is a personal thing. One auditor may feel confident whereas another may not feel confident. In adopting such an approach, one will need to employ the same auditors to audit the same organizations. This may not provide adequate career progression for some auditors.

Audit areas

These should be approached in the same way as described in Chapter 4. If adopting the extended assessment strategy, the samples need to be carefully selected to ensure that the areas, products, personnel and processes examined were not covered previously.

Approach to surveillance

Once a routine has been established, auditors can be tempted to make assumptions about the company's performance. They may take for granted that activities are proceeding to plan, fail to follow up a line of enquiry because they trust the management etc. Some of this is to be expected, otherwise the company will regard the auditor as a policeman. On other occasions the auditor may be frustrated by attempts to get the company to take corrective action and bend over backwards to help them.

The auditor has to remain alert to changes in the system and so must be vigilant at all times. If the auditor operates at a low level, confining enquiries to routines rather than the strategic issues, then he/she will be perceived as a plodder. Being accepted by top management and workers alike is one of the auditor's goals but to secure top management confidence the auditor has to talk their language. Several ways exist to achieve this:

• Become aware of news items that are relevant to the company's products and services.

• Become aware of changes in the industry that may affect the company.

• Become aware of emerging new standards or changes to the standards.

• Focus on strategic issues such as design, production and distribution strategies.

• Focus on cost effectiveness and encourage the use of quality costing techniques.

• Focus on performance improvement through use of the quality tools such as pareto analysis, design of experiments, QFD etc.

The auditor does not have to spend all the time auditing. Much can be gained from informal conversation, providing specific solutions to problems are not offered. The intention always should be to make the event one of added value to the company rather than a tour of the site followed by a free lunch. Trust on both sides can be built up through informal and formal communication with company staff. Show that you care about them, that they are not just any company but one which deserves your best efforts. Try to help them, not criticize them, and don't forget that the quality system is a tool of management. Helping to make this tool an effective instrument in achieving their purpose and mission will bring its rewards.

Reporting surveillance activities

The reporting of the audit findings should be carried out in the same manner as described in Chapter 5. The only additional item will be a report on the status of outstanding nonconformities: which were closed and which remain open.

Updating records

On completion of the surveillance visit the auditor's records should be updated. This should involve:

• Updating the audit programme, indicating when the audit was conducted

- Updating the corrective action, indicating which are open, which are closed and which have been allocated new completion dates

- Updating the trend analysis, indicating any new trends revealed

- Updating check lists with new items

Once complete, tracking corrective actions continues until the next surveillance visit, when the cycle is repeated.

Summary

This chapter has addressed surveillance audits, audits carried out to verify the quality system is being maintained. The following are the key issues covered:

◻ Surveillance is planned, conducted and reported in the same way as an initial audit.

◻ Surveillance is the most important attribute of quality system certification schemes.

◻ Several strategies exist for conducting surveillance: periodic assessment, extended assessment and continual assessment.

◻ Surveillance frequency depends on previous results, commitment, complexity and location.

◻ Selection of surveillance auditors depends on client/auditor relationships.

◻ Surveillance enables performance trends to be monitored.

◻ Corrective action proposals need to be evaluated for their effectiveness in preventing recurrence of nonconformities.

◻ Surveillance audits should address the maintenance and improvement elements of the system.

◻ Surveillance can degenerate into a formality or a value added event that builds trust between auditor and auditee.

Chapter 7

Requirements and their purpose

Introduction

This chapter deals with the requirements of ISO 9001. The purpose of the requirements in each element of the standard is given, together with the purpose behind the subject of each clause. The phrasing of some of the 'purpose statements' may appear somewhat negative since they all serve to *prevent* something. However, the purpose of the standard is preventive and one can either phrase the statements in this way or phrase them in a positive way: e.g. 'The purpose of these requirements is to ensure that something does not occur or cause something to occur.' The effect is the same. The purpose statements should assist auditors explain the need for action where nonconformities are detected and should assist in determining system effectiveness.

The applicability of each clause is given and where tailoring or interpretation is needed for specific industries, suggestions are provided. A summary of where tailoring is required is provided in Table 7-1.

ISO 9001 ELEMENT	4.1	4.2	4.3	4.4	4.5	4.6	4.7	4.8	4.9	4.10	4.11	4.12	4.13	4.14	4.15	4.16	4.17	4.18	4.19	4.20
Tailoring not required	X	X			X	X	X							X		X	X	X		X
Tailoring required			X	X				X	X	X	X	X	X		X				X	

Table 7-1 ISO 9001 tailoring matrix

In Appendix C the requirements of each clause are summarized and numbered so as to indicate how many requirements there are in the standard, in each element and in each clause. When asked how many requirements there are in ISO 9001, many would say 20. There are 20 elements but each element contains a number of sections, paragraphs and sentences. Where there is a 'shall' there is at least one requirement but often more than one (see box below).

REQUIREMENT DECOMPOSITION

Element	4.1 Management responsibility
Clause	4.1.1 Quality policy
'Shall' statement	The supplier's management with executive responsibilty shall define and document its policy for quality, including objectives for quality and its commitment to quality
Requirement 1	The supplier's management
Requirement 2	with executive responsibilty
Requirement 3	define policy for quality
Requirement 4	document policy for quality
Requirement 5	define objectives for quality
Requirement 6	document objectives for quality
Requirement 7	define commitment to quality
Requirement 8	document commitment to quality

In this example there are eight different requirements. Whilst such decomposition may appear rather extreme it is quite possible for instance that:

• The policy has been defined but not by the supplier's management.

• The supplier's management have defined the policy but not those managers with executive responsibility.

• The commitment has been defined but not documented.

• The policy has been documented but not the objectives, etc.

Another example is in the requirement 'The supplier shall establish and maintain documented procedures to control all documents and data that relate to the requirements of this International Standard' there are six requirements. Everywhere there is an 'and' there is an additional requirement, so dissecting this requirement creates six separate requirements as illustrated below:

- One for establishing documented procedures for controlling documents
- One for maintaining documented procedures for controlling documents
- One for establishing documented procedures for controlling data
- One for maintaining documented procedures for controlling data
- One for controlling documents that relate to the standard
- One for controlling data that relates to the standard

There are in fact several more if we treat the requirement for 'documented procedures' as two requirements: one for procedures and one for documents. One of the most common deficiencies is that suppliers fail to take into account all of the individual requirements in the standard. They may have established a procedure but not documented it. They may have established and documented a procedure but limited it to the control of documents, ignoring data controls. They may have done all of these things and omitted to control a specific document that relates to the requirements of the standard. When auditors detect nonconformities, they detect that one of these individual requirements has not been met. In this chapter, 323 individual requirements have been extracted from the wording of the standard. Often, requirements to establish and maintain are treated as one requirement rather than two, so 323 is not the total, it is more likely to be closer to 500. This list clearly demonstrates that an ISO 9001 audit is no simple task. Even if we were to take only one sample to test conformity, it will take considerably longer than most certification bodies allow to verify conformity with *all* requirements of the standard. Many requirements can be verified in seconds, as one is shown a documented procedure and evidence that it is in use. However, many other requirements will take considerably longer to confirm conformity, such as the requirement that results of inspections and tests meet specified requirements. How could an auditor know this unless a detailed examination was carried out on a sample. This means that auditors need to examine the plans, specifications and records to verify conformity, not merely verify that such documents exist and that they *appear* under control.

In the lists provided, the following are treated as addressing one process or item rather than several (otherwise the lists would seem repetitive):

- Inspection and test, since the term 'verification' could have been used

- Define and document, since one cannot reasonably document something without defining it to some degree

- Identify and document for the same reasons as above

- Inspection, measuring and test equipment, since these could have been referred to as measuring devices

- Production, installation and servicing, since these are each operating processes

An important principle to remember when using ISO 9001 is that it only contains a minimum set of requirements for obtaining assurance that products will meet specified requirements. It does not address all operations of a business. Auditors do not need to verify that every requirement of internal procedures is implemented since many of these requirements will serve purposes other than that of ensuring product meets specified requirements.

A detailed analysis and interpretation of these requirements is provided in the *ISO 9000 Quality Systems Handbook*[1].

4.1 Management responsibility

The purpose of these requirements is to prevent the organization deviating from its policies and objectives and retreating from its commitment to quality. It also prevents employees from performing activities and making decisions for which they are not authorized or trained.

4.1.1 Quality Policy

The purpose of defining the quality policy is to direct everyone in the organization in a particular direction regarding quality and give them a sound basis for the actions and decisions they execute. The quality system implements the quality policy and therefore the quality policy establishes the requirements that govern the scope and applicability of the quality system.

The requirements apply to the organization's top management. No tailoring is needed for specific industries.

---

[1] ISBN 0 7506 2130 3

4.1.2 Organization

4.1.2.1 Responsibility and authority

The purpose of defining responsibility and authority of personnel is to ensure that staff are in no doubt as to the results they are expected to achieve, the decisions they have a right to take and to whom they are accountable for their actions and decisions.

The requirements apply to all the organization's managers responsible for activities affecting the quality of the products and services provided to external customers. No tailoring is needed for specific industries.

4.1.2.2 Resources

The purpose of identifying resource requirements is to ensure that the organization is equipped with the capability needed to honour its intentions and its commitments.

The requirements apply to the organization's management. No tailoring is needed for specific industries.

4.1.2.3 Management Representative

The purpose of appointing a Management Representative is to ensure that quality system development, implementation and evaluation is planned, organized and controlled with the delegated authority of the organization's executive management.

The requirements apply to the organization's top management. No tailoring is needed for specific industries.

4.1.3 Management review

The purpose of management review is for executive management to establish whether the quality system as a whole remains suitable and effective for its intended purpose and, if not, to direct it be changed.

The requirements apply to the organization's top management. No tailoring is needed for specific industries.

4.2 Quality system

4.2.1 General (quality system development)

The purpose of establishing a quality system is to provide a management tool to implement the quality policy and ensure that all products and services supplied to customers satisfy their needs and expectations.

The purpose of these requirements is to prevent an unstructured and incoherent approach to satisfying the requirements of the standard both in general and for specific products, projects or contracts.

The requirements apply to the organization's management. No tailoring is needed for specific industries.

4.2.2 Quality system procedures

The purpose of quality system procedures is to implement policy and regulate processes that are essential to the business.

The requirements apply to the organization's management. No tailoring is needed for specific industries.

4.2.3 Quality planning

The purpose of quality planning is to determine the provisions required to ensure that particular requirements for products and services will be achieved by the organization and to identify where the provisions of the existing quality system may need to be changed.

The requirements apply to the organization's management responsible for executing orders, contracts or projects. They apply where the quality system has to be tailored to meet customer requirements. They only apply to product or service quality, not to internal quality improvement programmes. These are covered (indirectly) by the corrective and preventive action requirements.

4.3 Contract review

The purpose of these requirements is to prevent the organization from entering into a commitment with its customers that it may fail to meet.

The requirements apply to the organization's management with responsibility for order input. Tailoring is necessary for specific industries.

Where an organization does not operate through formal customers' orders or contracts the requirements should be interpreted as applying to the transaction between the organization and its 'customers', i.e. the people who use its services and the people or authorities who dictate what services will be provided. Such transactions may be verbal or written. Often the person in the organization receiving the request takes the responsibility for defining and documenting it.

4.4 Design control

The purpose of these requirements is to prevent design commencing without agreed requirements and to prevent the release of designs that do not meet user needs and/or requirements.

The requirements apply to the core products or services designed by the organization for sale to its customers (its outputs) and not to any support or ancillary products that are not offered for sale, such as tools and test equipment or internal services. All the requirements of element 4.4 apply to the same core products and services and should not be indiscriminately applied to internal outputs of the organization. Tailoring is necessary for specific industries

4.4.1 General (design control procedures)

The design control procedures required should cover all the requirements of element 4.4. Even though each clause does not specifically require procedures, there is an implication that to control design the procedures need to address each of the requirements. There could be one procedure to cover the complete process or several procedures, either one for each clause (which would be unusual) or a number of procedures that address the various stages and activities of the design process. Separate procedures may be necessary for hardware, software and service design. Some may apply to all three types of product, others may be applicable only to one type of product.

4.4.2 Design and development planning

The purpose of these requirements is to provide a means of defining the activities needed to produce, verify and validate a design that meets the specified requirements.

4.4.3 Organizational and technical interfaces

The purpose of these requirements is to communicate organizational and technical requirements to those assigned to execute them. Organizational interfaces refer to the relationships between organizations, both internal and external to the organization. Technical interfaces refer to the relationships between components within the end product and between the end product and those with which it will interface in service.

4.4.4 Design input

The purpose of these requirements is to provide a means of establishing design requirements for determining the acceptability of the design solution. The design input requirements may be specified in a contract or other agreement or in standards and other documents referenced in the contract or agreement.

4.4.5 Design output

The purpose of these requirements is to provide a means for describing the design solution and preventing its release before its acceptability has been established.

4.4.6 Design review

The purpose of these requirements is to prevent design continuing beyond prescribed stages before its acceptability has been established. Design review is a method of design verification.

4.4.7 Design verification

The purpose of these requirements is to provide a means of establishing objective evidence that the design output meets the design input. Design verification may not be possible for services until they are in use, unless modelling is employed to test theories.

4.4.8 Design validation

The purpose of these requirements is to provide a means for confirming that the design reflects a product which conforms to user needs and/or requirements. In the service sector, design validation may be synonymous with design verification.

4.4.9 Design changes

The purpose of these requirements is to provide a means for ensuring that only approved design changes are implemented.

4.5 Document and data control

The purpose of these requirements is to prevent the inadvertent use of incorrect, invalid or obsolete documents and data.

These requirements apply to all documents and data that relate to the requirements of the standard, not just the quality system procedures. Such document and data include policies, procedures and documents derived from such policies and procedures which prescribe requirements essential to the achievement of product and service quality. Data may be in the form of information contained in documents or in electronic storage such as a database on a computer. They do not apply to documents and data which record the results of activities performed or results achieved such as test results. The requirements also do not apply to the activities performed prior to submission of documents and data for review and approval. No tailoring is needed for specific industries.

4.5.1 General (document and data control procedures)

Document control procedures are required for all documents and data that relate to the requirements of the standard; therefore this may require more than one control procedure. The auditor needs to establish what types of documents and data are employed by the organization and ensure the control procedures cover all such types identified, including external documents such as standards and manuals whether supplied by customers, suppliers or obtained from other sources.

4.5.2 Document and data approval and issue

The purpose of these requirements is to provide a means for preventing invalid documents and data being issued or selected for use.

4.5.3 Document and data changes

The purpose of these requirements is to provide a means for preventing unauthorized changes to approved documents and data, of ensuring that those authorizing the documents make sound decisions and those using the changed documents know what has changed.

4.6 Purchasing

The purpose of these requirements is to prevent the purchase and acceptance of product or service that does not meet specified requirements.

These requirements apply to products and services purchased from external suppliers for use in connection with the products and services offered for sale to external customers. They do not apply to purchases that have no bearing on the saleable product such as stationery, word processors, furniture, vehicles and food unless such items form part of the saleable product or service or are used in a way that would affect it. No tailoring is needed for specific industries.

4.6.1 General (purchasing procedures)

The requirement for purchasing procedures applies only to purchased product that form parts of or are used in saleable products or services. They also apply to the purchasing process wherever that process commences, be it in design, in production or elsewhere. They do not only apply to the purchasing function.

4.6.2 Evaluation of subcontractors

The purpose of these requirements is to prevent the use of external suppliers for which there is no objective evidence of their ability to meet the purchaser's requirements.

4.6.3 Purchasing data

The purpose of these requirements is to provide a means of communicating purchasing requirements to subcontractors and for establishing the acceptability of supplies provided by the same subcontractors.

4.6.4 Verification of purchased product

4.6.4.1 Supplier verification at subcontractor's premises

The purpose of these requirements is to prevent subcontractors prohibiting the purchaser from gaining access to their premises to carry out legitimate verification activities. They only apply when confidence in supplies cannot be obtained solely by remote means.

4.6.4.2 Customer verification of subcontracted product

The purpose of these requirements is to provide a means for the purchaser's customer to verify purchased product prior to shipment to the supplier. They only apply in cases where customers need added confidence in the supplier's control over its subcontractors. Such instances will be specified in the contract with the customer when they are necessary.

4.7 Customer supplied product

The purpose of these requirements is to prevent loss, damage or deterioration of product not owned by the organization and to prevent the use of any such product that is unsuitable.

These requirements apply to product furnished by the external customer either for incorporation into product that will be subsequently sold back to the customer or for use in executing the contract and which will be subsequently returned to the customer following completion of the contract. Not all the requirements can be applied to customer supplied services. Product may be tangible goods or may be information such as personal details provided to banks, credit card agencies etc. which have to be protected. In education, training, health care etc. the customer supplied product is the pupil, student, delegate, patient etc. In a law practice or financial institution it is any document furnished by the client which will ultimately be returned.

4.8 Product identification and traceability

The purpose of these requirements is to prevent inadvertent use or installation of incorrect product and to enable the recurrence of nonconformities to be prevented.

The identification requirements apply to products or services supplied to customers after receipt of the component parts or materials and through to delivery and installation. They do apply to support products and materials etc. used to produce the deliverable product where absence of identification would cause product nonconformity. They apply only where products cannot be distinguished from one another without additional markings or labels.

The traceability requirements apply only when is it necessary to recall product following the identification of defects and only apply to the traceability of the product and not to the activities carried out on it such as tests, heat treatment etc. Other requirements in clause 4.14.2 imply a need for traceability in order to establish the cause of nonconformity. Tailoring is necessary for specific industries.

4.9 Process control

The purpose of these requirements is to prevent variation in result producing processes from exceeding defined limits.

These requirements apply to production, installation and servicing processes and not to the supporting processes or other processes dealt with in other elements of the standard, such as design, purchasing and delivery. In the service sector they apply to the processes needed to deliver the service, whether activated by a specific customer or not. Tailoring is necessary for specific industries.

4.10 Inspection and testing

The purpose of these requirements is to prevent the release of nonconforming product to a subsequent stage in the process.

Whilst entitled 'Inspection and testing' these requirements apply to all verification activities carried out to determine whether products and services and parts thereof comply with specified requirements. The verification method can be an analysis, a simulation or a comparison. Observing services during their delivery is a verification, as is a direct examination of the results produced by the services. Although final inspection implies an inspection carried out prior to delivery, the point of delivery can be on completion of a service to establish that it was executed properly. Tailoring is necessary for specific industries

4.10.1 General (inspection and test requirements)

These requirements apply to the process of generating the inspection and test requirements. This process may be part of a design, development or planning process rather than an inspection process. Many such requirements may be included in the manufacturing drawings, specifications or quality plan but they may also be supplemented with additional requirements for sampling etc. These documents would define the tests to be carried out rather than how the test should be performed. They apply also to the procedures used for controlling inspection and test activities, as well as those used for performing specific inspections and tests: e.g. how a specific test would be performed using specified test equipment, facilities and tools. Whilst clause 4.10.5 covers records, the requirements of this clause dictate that the inspection and test documents include requirements for records that will contain the results.

4.10.2 Receiving inspection

These requirements apply to product received by the supplier but may be satisfied before receipt. Just-in-time techniques are therefore acceptable providing there is evidence justifying the confidence the supplier has to omit inspection on receipt.

4.10.3 In-process inspection and testing

These requirements apply at any stage between receipt and final inspection.

4.10.4 Final inspection and testing

These requirements apply to the last inspection and test stage for a particular product. This may be prior to delivery to an internal stock room, dispatch to a customer or, where responsibilities extend to installation, final inspection will be prior to handover to the customer. There may be several final inspections before ownership of the product passes to the customer.

4.10.5 Inspection and test records

These requirements apply to all the quality records identified in the inspection and test requirements generated to meet clause 4.10.1.

4.11 Inspection, measuring and test equipment

The purpose of these requirements is to prevent devices being used for product, service and process verification that will not yield accurate and precise results.

Although the heading implies tangible equipment only, these requirements apply to any device used to determine the acceptability of products and services. The device may be an item of equipment but can also be a piece of software or a reference material when measuring quantitative characteristics. The device can also be a technique such as a written examination or a questionnaire when measuring achievement of qualitative characteristics. Clearly in such circumstances many of the requirements of this element could not be applied to non-physical measurement techniques and would therefore be not applicable. In applying these requirements it is useful to substitute the term 'measuring devices' for inspection, measuring and test equipment. Tailoring is necessary for specific industries.

4.11.1 General (measurement control, calibration and maintenance requirements)

These requirements specify general requirements, some of which are amplified further in clause 4.11.2. Hence the general requirement for control, calibration and maintenance is in 4.11.1 and detail requirements in 4.11.2.

4.11.2 Control procedure

It is not intended that all these requirements be satisfied by one procedure. Several may be needed. Some of the requirements relate to the process where parameters to be measured are identified. This may be during design, development or planning rather than in the calibration process which deals with particular measuring devices. Certainly in meeting the requirements in clause 4.10.1 the supplier will implement the first two of the following requirements.

4.12 Inspection and test status

The purpose of these requirements is to prevent inadvertent use, dispatch or installation of product that has not been verified or that has failed to meet prescribed verification requirements.

These requirements apply to any product or service that is subject to verification and could be termed 'verification status'. In the service sector, verification status applies to whether a service is operational or non-operational. Such status is often exhibited by notices such as the 'Out of Order' notice on an automatic cash dispenser in a bank. Tailoring is necessary for specific industries.

4.13 Control of nonconforming product

The purpose of these requirements is to prevent the unintended use of product that does not conform to specified requirements.

These requirements apply to any product or service that has been subject to verification and found not to conform to specified requirements. They only apply to the particular product or service found nonconforming and those exhibiting the same deficiency, whether before or after delivery of product or service. They do not apply to processes or quality systems. They also do not apply to product not yet produced or to the action taken to prevent recurrence of the nonconformity. The action taken to correct a nonconformity is termed correction[2], not corrective action. A better term would be remedial action. Tailoring is necessary for specific industries.

4.13.1 General (nonconformity control procedures)

These requirements apply to the procedures that regulate the disposal of nonconforming product or service. The procedures should cover all the aspects addressed in the requirement including those covered in clause 4.13.2. There may be one procedure or several, depending on the type of product and where in the life cycle the nonconformity is detected. Hence they apply after a product or service has been inspected and not before. They do not apply to product that has been released to customers. This condition is addressed in clause 4.14.2.

4.13.2 Nonconformity review and disposition

These requirements apply to the disposal of nonconforming product or service including any which is deemed usable without change, i.e. fit for purpose but not fully meeting the requirement.

4.14 Corrective and preventive action

The purpose of these requirements is to prevent the recurrence of identified nonconformities and the occurrence of potential nonconformities.

The corrective action requirements apply to any product, service, process or quality system possessing nonconformities, and do not solely apply to products. The preventive action requirements apply to all products, services and operations affecting the quality of those products and services. Corrective actions can only be taken when there has been a nonconformity. Preventive action can only be taken when a potential nonconformity has been identified. No tailoring is needed for specific industries.

4.14.1 General (corrective and preventive action requirements)

These requirements apply to the actions necessary to correct and prevent nonconformity in products, services, processes and systems, not just products, whereas element 4.13 only applies to products and services.

---

[2] See ISO 8402 1994

The Tale of the Lonely Auditor

Amanda was to meet Paul, the Marketing Manager, at 9 o'clock and was guided there by Julie. Paul greeted Amanda and asked her to sit down but Amanda noticed that there was no chair for her guide. It was clear to Amanda that the office was a little too small for the interview so she asked Paul if there was a more suitable room they could use. Paul discovered that the conference room was free until 10 o'clock.

On arrival at the conference room Amanda noticed that there were no windows but proceeded with her interview anyway. Paul asked Amanda if she would like coffee and Julie volunteered to get it to save time.

Whilst Julie was organizing the coffee, Amanda asked Paul to explain the Marketing process from beginning to end. Some 30 minutes later, Julie arrived with the coffee apologizing that she had been diverted to resolve a problem with a customer.

Amanda latched on to this incident and asked to examine the file of customer complaints. Paul left the room to locate the customer complaints file. Some 10 minutes later, Paul returned to explain that as the file had been in use, there had been a problem locating it. Amanda examined the file and picked upon one compliant to probe the corrective action. The evidence Amanda needed was not in the file so Paul again left the room. This time he did not return, so Julie volunteered to look for him.

It was now 10 a.m. and the conference room was required by a Sales team so Amanda had to leave. Not having either a guide or the manager she was interviewing, she was at a loss as to what to do next - stay put or seek them out ... but where had they gone? She had allocated one hour for this interview and had progressed very little apart from suspecting that the company might be having problems satisfying its customers. There was nothing concrete on which a nonconformity statement could be based ... or was there? She still had the file of customer complaints in her possession!

The moral of this story is:

Don't let them out of your sight for a moment - go with them!

4.14.2 Corrective action

These requirements apply to the procedures employed for eliminating the cause of nonconforming products, services, processes or quality systems. There may be one procedure or several depending on the variety of situations for which corrective action is needed. The procedures would not normally include the method of changing documentation since this aspect would be addressed by the document control procedures produced to satisfy clause 4.5.1. The requirements also apply to external reports of nonconformity in products and quality systems from customers or third parties.

4.14.3 Preventive action

These requirements apply to the procedures employed to prevent the occurrence of nonconformity in products, services, processes and quality systems. Since nonconformity prevention is often achieved by careful planning, these requirements are not a substitute for the quality planning requirements of clause 4.2.3. However, some of the techniques used for planning purposes can also be used to satisfy this requirement, such as fault tree analysis, failure modes and effects analysis, and quality costing. Other techniques, though, can only be used once data exists, such as Pareto analysis.

4.15 Handling, storage, packaging, preservation and delivery

The purpose of these requirements is to prevent damage or deterioration to conforming product.

These requirements apply to conforming product received or produced by the supplier. They also apply to unserviceable and nonconforming products so as to prevent further damage and deterioration. In the service sector they would apply if the service could be prone to damage or deterioration by storage, handling etc. A client may provide transport, packing cases, luggage, letters etc. for the supplier to use or deliver. The requirements can be applied to any service that contains a product such as data, documentation, plastic cards, disks or any tangible item. In education and training the requirements do not apply to the people being educated or trained but to the material, knowledge and skill being delivered to them. The requirements of clause 4.7 apply in such cases. Tailoring is necessary for specific industries.

4.15.1 General (conformity control procedures)

These requirements apply to the procedures to be employed to safeguard product or service and may be product specific or general. One of the anomalies of the requirement is that a procedure for delivery of product is required but on reading the specific delivery requirement it will be seen that it only applies to safeguarding the product and not to the delivery process.

4.15.2 Handling

This requirement applies to handling of product at any stage in its life cycle whilst it is under the supplier's control and not just the finished product.

4.15.3 Storage

These requirements apply to any area in which product is stored whether awaiting receipt inspection, processing or delivery. Such areas need to be designated as storage areas so that adequate controls are employed.

4.15.4 Packaging

These requirements apply to any packaging applied to product on receipt, prior to or after processing or for transportation to customers. They also apply to any marking processes whether applied to the packaging or the product. Packing applies to the act of packing items whereas packaging applies to the materials employed in packing an item.

4.15.5 Preservation

These requirements apply to preservation methods employed when product is in storage, in-process or in transit. Segregation, whilst not included in the title, is a means of preservation, i.e. it prevents cross-contamination and inadvertent misuse.

4.15.6 Delivery

As stated above, these requirements only apply to protection of product after final inspection rather than the delivery process as a whole. Delivery may take place between locations under the supplier's control as well as to locations designated by the customer.

4.16 Control of quality records

The purpose of these requirements is to prevent documentary evidence of conformity with specified requirements from being irretrievable.

These requirements apply to all records classified as quality records. Records are historical documents; they record events that have taken place. Quality records are the output from processes and constitute the results of formal product, service, process and system verification activities. They provide objective evidence of the achieved characteristics of products and services and the results of system or supplier assessments, audits, examination etc. Not every record is a quality record, only those identified in ISO 9001 by a cross reference to clause 4.16 of the standard. No tailoring is needed for specific industries.

4.17 Internal quality audits

The purpose of these requirements is to prevent a deterioration in agreed standards.

These requirements apply to all operations of the organization governed by the quality system. They do not apply to external audits, whether second or third party audits. They apply also to product audits, process audits and service audits as well as quality system audits, as evident form the definition given in ISO 8402. No tailoring is needed for specific industries.

4.18 Training

The purpose of these requirements is to prevent unqualified personnel from performing activities that affect quality.

These requirements apply to all personnel, whether employees or contract labour, who execute activities governed by the quality system. They do not apply to the training of customers or suppliers except where such personnel are working under the organization's quality system. No tailoring is needed for specific industries.

4.19 Servicing

The purpose of these requirements is to prevent servicing being carried out under uncontrolled conditions.

These requirements only apply when the customer-supplier agreement includes post delivery support whether for products or services (N.B. Servicing is not the same as Services. Servicing is what is done to something; Services are what is provided.) In such cases they apply to the servicing of products or services designed, produced or delivered by suppliers. They relate to the services carried out after delivery of the product or service, such as warranty repairs, maintenance, technical support etc. In a bank, customer care programmes, keeping customers appraised of latest interest rates etc., is servicing. In a hospital, post-operative care is servicing. Tailoring is necessary for specific industries.

4.20 Statistical techniques

The purpose of these requirements is to prevent invalid techniques being used to establish, control or verify product or process capability.

These requirements apply to product and process acceptance decisions taken on the basis of data generated by statistical means. They do not apply to the manipulation of data by methods that do not rely on probability theory. No tailoring is necessary for specific industries where the techniques are employed.

4.20.1 Identification of need

These requirements apply to the process used to identify whether statistical techniques are needed to accept products and the processes that produce the product.

4.20.2 Procedures

These requirements only apply to the statistical techniques identified when meeting clause 4.20.1. Procedures are not necessary for other uses of statistics.

Chapter 8

Inconsistencies in ISO 9001

Introduction

In this chapter, the requirements of ISO 9001 are analysed and the inconsistencies revealed to help auditors avoid the pitfalls that can arise when making assumptions about the standard. Arguments are provided showing how these inconsistencies may be overcome but these should not be taken as official interpretations. Many certification bodies have their own interpretations of the ISO 9001 and these should be used. However, it is hoped that interpretations given herein may be helpful in certain situations.

Consistency is a quality characteristic of a document. It is determined by the extent to which a document unifies communication both within itself and between other documents with which it is related. The standard defines minimum requirements for a quality system to give an assurance of quality and therefore it must be assumed that where the wording of particular clauses appears inconsistent with other clauses or other related documents, this difference in wording is intentional so as not to impose too onerous a requirement. If the inconsistencies were not intentional then the standard ceases to be a standard and of little value as a means of measurement. It may be of immense value as a guide but for measurement purposes it would be unfit for its purpose.

Inconsistencies therefore arise when a document does not convey the messages that were intended. Suppliers may not read into the standard the many implications that requirements intend to convey and so may challenge auditors to justify their nonconformity statements. In some situations the challenge may be sustained because it cannot be proven that product quality will be, is being or has been adversely affected.

Such inconsistencies in ISO 9001 may arise where there is, for example:

- An omission in a requirement due to imprecise wording

- An implication but not a clearly specified requirement

- A duplication of a requirement i.e. two or more clauses address the same requirement but with alternative wording

- A requirement in one clause for something to be defined and documented and another clause where it is only to be defined, implying that documentation is not required

- A requirement in one clause for procedures or records to contain or include certain aspects but not in other clauses

- A requirement for something to be stipulated or specified but not defined and documented as in other clauses

- A requirement that gives the reason in some clauses but not in others

- An ambiguity in the wording creating uncertainty in what is actually required

- A requirement for documented procedures and records in some clauses but not in others

Application of the standard

The ISO 9001 standard is stated as specifying quality system requirements for use where a supplier's capability to design and supply conforming product needs to be demonstrated. It is also stated that the standard applies in situations where design is required and the product requirements are stated in performance terms. However, in both cases these statements imply that it is not intended that the standard be used where such capability does not need to be demonstrated or when design is not required. It is unclear as to whether the organization requiring design capability to be demonstrated and requiring product to be designed is intended to be the customer, the user or the supplier. If it is the supplier then ISO 9001 applies to any organization that carries out design. If it is the customer or user then ISO 9001 applies only where they require product to be designed to meet their specific requirements. This is a common source of misunderstanding. If a supplier applies for certification against ISO 9001 and due to the concentration of nonconformities in design requests re-certification against ISO 9002, the auditor cannot grant such a request. This is because the application for certification against ISO 9001 was based on the fact that the supplier is required to carry out design by their customers. They would not be qualified to tender for design contracts if they only possessed ISO 9002 certification.

Clause structure

If a major nonconformity is one where the requirements of a clause have not been met and a minor is where one requirement of a clause has not been met, then a failure to maintain management review records is a minor nonconformity since it is only one requirement of a clause, whereas a failure to maintain contact review records is a major nonconformity because this is given a separate clause number. However, in the 1987 version of the standard both would have been minor nonconformities. Clearly the standard was not structured so as to facilitate the use of these definitions of nonconformity.

Quality policy

Policies including objectives

There is an ambiguity in the quality policy statement as it is unclear whether it should include objectives for quality or whether the quality objectives can be defined and documented separately from the quality policy. Whether or not the objectives are included in the policy, they need to be separately defined and documented. Policies are not objectives and vice versa[1]. Separate statements of policy and objectives should not be considered a nonconformity.

Relevance of objectives to goals and expectations

There is no requirement for the quality objectives to be relevant to the supplier's organizational goals and expectations and needs of customers if they do not form part of the quality policy statement. As it would be illogical for the objectives not to be as relevant to these goals and the policy the requirement is implicit. Objectives that do not align with the organizational goals and expectations and needs of customers are not serving the organization or its customers and therefore show a lack of commitment to quality.

---

[1] Refer to *ISO 9000 Quality Systems Handbook* for further explanation.

Objectives being understood, implemented and maintained

Quality policy has to be understood, implemented and maintained but not the quality objectives if they do not form part of the quality policy statement. There would be no benefit to the organization in defining and documenting its quality objectives if it did not intend to implement them, maintain them and ensure their staff understood them. Organizations that ignore quality objectives would be showing a lack of commitment to quality. An alternative argument is to refer to clause 4.2.3 where the supplier is required to document how the requirements for quality will be met. The definition in ISO 8402 for 'requirements for quality' relates the expression of needs to an entity which is defined as relating to an organization as well as a product, process etc. It follows therefore that quality objectives are an expression of the needs of the organization and therefore clause 4.2.3 requires that the means by which these objective be met be documented: e.g. a quality plan for achieving the organization's quality objectives. This is a rather convoluted argument but it is logical even though this may not have been the original intent of clause 4.2.3.

Procedures for quality policy and objectives

Documented procedures are not required for establishing and maintaining the quality policy and quality objectives and for ensuring that they are understood and implemented. Documented procedures themselves are not essential even if the argument above referring to clause 4.2.3 is used, as no documented procedures are required in this clause. However, clause 4.2.1 requires the quality system to be documented and the Quality Manual to cover the requirements of the standard; therefore the means by which the policy and objectives are established and maintained, understood and implemented should be defined in the Quality Manual.

Responsibility and authority

Procedures for responsibility and authority

Documented procedures are not required for the assignment, review and evaluation of responsibility and authority. Responsibilities and authority are required to be documented in clause 4.1.2.1. This documentation forms part of the documented quality system that is subject to audit. Clause 4.17 requires a documented procedure for internal quality audit; therefore as part of the audit process, responsibilities and authority of personnel should be reviewed and evaluated for their effectiveness. However, there are no criteria in clause 4.1.2.1 for the assignment of responsibility and authority; the supplier has to formulate its own rules. If there are conflicting responsibilities or authority, duplication or gaps then the absence of such rules indicates that a potential cause of nonconformity exists for which preventive action is required under clause 4.14.3. The nonconformity would be that the preventive action procedures make no provision for responsibilities and authority to be reviewed to detect, analyse and eliminate potential causes of nonconformity.

Clearly defined responsibilities

Responsibilities are not required to be *clearly* defined and documented. If this situation arises then the argument above can be used to secure corrective action.

Resources

Documentation of resources

Resources have to be identified but not defined and documented, unlike many other subjects of the standard. The auditor should refer the supplier to clause 4.2.1, which requires the system to be documented; therefore, whatever the method used to identify resources, the resources that have been identified have to be documented.

Qualified personnel

The standard requires the assignment of trained personnel but not qualified personnel as specified in clause 4.4.2 for design and development, or educated and experienced personnel as specified in clause 4.18. The term used should have been 'competent personnel'. There is no requirement in this clause for the personnel to be trained in the task that has been assigned to them. Providing they have had some training, no matter how irrelevant, the supplier would be compliant. However, an auditor may take a different view. The auditor should not insist that all personnel be trained irrespective of this requirement. Personnel need to be competent and a test of competence is the ability to perform the job effectively over a long period of time.

Documented resource procedures

Documented procedures are not required for ensuring that adequate resources are identified acquired and allocated. Auditors cannot require documented procedures for these activities; however, clause 4.2.1 does require the system to be documented therefore the system for identifying and providing resources should be documented in the system.

Resource records

Records of the specific resources provided do not have to be maintained. Alas there is no way an auditor can require such records to be produced unless ISO 9001 is a contract requirement. In such cases, clause 4.16 requires records to be maintained to demonstrate conformance to specified requirements and therefore can be quoted.

Management Representative

Executive responsibility

The Management Representative (MR) is not required to have executive responsibility. The MR is required to have authority to implement and maintain the quality system which must bring with it the authority to resource the activities defined in the quality system. Only executive managers are usually given authority for resources. If the MR does not have authority for resourcing then it needs to be demonstrated how he/she ensures (i.e. causes) the system to be implemented and maintained. If the MR works through persuasion and in cases of conflict refers to the CEO, then clearly the MR does not have the required authority.

Management review

Quality system review

The title is not wholly consistent with the requirements of this clause as the review required is only of the quality system and not of any other aspect of management. The activity can be defined as a management review, i.e. a review by management or a quality system review.

Single manager review

If the Management Representative does have executive responsibility, there is no nonconformity if the quality system is only reviewed by the Management Representative. ISO 9004 recommends that the review of the system be independent and be performed by top management. Clearly the MR is not independent and if there are other executive managers then they should be involved. It is no nonconformity if the MR is the CEO and there are no other executive managers but is worthy of an observation that the assignment of responsibilities could be more effective. But in other cases the term 'management' means those who represent the management and this includes all of them, so a single-man review would be a nonconformity.

Documented intervals of management review

The intervals of management review have to be defined but not documented. As the quality system is required to be documented in clause 4.2.1, the defined intervals have to be documented in the system.

Review of documented procedures

There is a requirement to review the quality system at defined intervals but not its documented procedures (clause 4.2.2b) or the other documents used in the quality system (clause 4.2.1). The quality system consists of procedures, responsibilities, resources, processes and organization structures. A review of the system includes a review of all of these, so the standard does not have to specifically require procedures to be reviewed. In this respect the auditor needs to establish when each of the procedures was last reviewed. An en-bloc review is insufficient evidence that consideration has been given to their suitability and effectiveness as required by clause 4.1.3.

Content of management review records

The content of management review records is not specified so can be a simple date log or a file of charts, histograms, tables etc. A quality record is the result of an activity that provides objective evidence of the fulfilment of the requirements for quality[2]. The activity is a review of the quality system; therefore the record should contain objective evidence of the suitability and effectiveness of the system in satisfying the requirements of the standard and the supplier's stated quality policy. The minutes of the management review meeting contain the results of the meeting, not the results of the quality system review. The meeting should be convened to discuss the results of the review so that decisions can be made. When records of management reviews do not provide such objective evidence the nonconformity is against ISO 8402, which is invoked in ISO 9001. ISO 8402 contains provisions of ISO 9001 as stated in clause 2 of the standard.

Quality system

Documenting the quality system procedures

There is a requirement for the quality system to be documented as well as procedures to be documented, thereby duplicating the requirement. It is possible for a supplier to have procedures that are not documented, hence the requirement for documented procedures in many classes of the standard. However, even without this requirement, procedures would have to be documented since they form part of the system that itself has to be documented.

---

[2] ISO 8402

Purpose of the quality system

The quality system has to provide a means of ensuring that product conforms to specified requirements and not a means for implementing the supplier's quality policy and objectives. This is an oversight, as logically the system should be the means of implementing the policy and achieving the objectives of the organization. If the system is found not to contain provisions for implementing the stated policy and objectives then it is ineffective and nonconforming with clause 4.1.3, as the organization has not ensured the continuing suitability and effectiveness of the system.

Scope of Quality Manual

The Quality Manual has to cover the requirements of the standard but is not required to respond to all requirements of the standard or the supplier's quality policy. Covering a requirement depends upon what a requirement is: an element, a clause or a 'shall' statement or part of such a statement. Where the manual does not cover the requirements to this extent, it should cross refer to the documents that do respond to these statements. If this is deficient then there is just cause for a nonconformity on the basis that the manual does not cover the requirements of the standard either directly or indirectly.

Outline of documentation structure

The Quality Manual has to outline the structure of the documentation used in the quality system but is not required to describe or specify the documentation structure. An outline, by definition, should include all types of documents, including the types of quality records, that constitute the quality system documentation, otherwise the outline is incomplete.

Changes to the quality system

There is no requirement that changes to the quality system have to be documented, reviewed and approved prior to implementation, unlike design changes in clause 4.4.9. Maintenance of the quality system is often interpreted as maintaining documents when in fact the system consists of much more than documentation, as indicated by the definition in ISO 8402[3]. System changes are not the same as documentation changes. The former changes the configuration of the system, whereas the latter only changes documentation. For a design change to be agreed, its effect on all components of the design needs to be considered and likewise, for a quality system change to be

---

[3] A quality system is defined in ISO 8402 as the organizational structure, procedure, processes and resources needed to implement quality management.

agreed, its effect on all components of the quality system needs to be considered. For a document change to be agreed only changes to that document need to be considered. Auditors should therefore establish how the *system* is maintained and not only how documentation is maintained (see *ISO 9000 Quality System Handbook*, Part 2, Chapter 2).

Quality system procedures

Constitution of procedures

There are requirements for quality system procedures rather than for all the documentation used in the system. The only procedures required are those that are required to be established and maintained by the various clauses of the standard. All other documents should be derived from the established procedures since no other provisions are included in ISO 9001.

Procedures for developing procedures

Documented procedures are not required to design and develop the quality system, only for controlling quality system documentation as specified in clause 4.5.1 once they exist. If it can be shown that the review and approval process produces documents of an inconsistent standard that reduces system effectiveness then a nonconformity exists with clause 4.5.2 addressing the adequacy of documentation.

References to quality records

Documented procedures are not required to reference the quality records that should be established and maintained to demonstrate conformance to specified requirements and the effective operation of the quality system. A system can therefore be compliant if records exist without any cross reference to procedures. There is no requirement for the quality records procedures to include compilation instructions. If it can be shown that the absence of compilation instructions or cross references to quality records causes errors in the implementation of the system then a nonconformity is justified on the basis that the procedures are not being effectively implemented as required by clause 4.2.2b.

Procedures beyond the scope of the standard

Procedures have to be prepared consistent with the requirements of the standard and *the supplier's quality policy* which conflicts with clause 4.2.1. This clause only requires the quality system to ensure that (intended) product conforms with specified requirements. If the quality policy imposes obligations beyond the scope of the standard – e.g. personnel safety and environmental issues – then procedures also have to go beyond the standard. If a supplier is found not to comply with a quality policy that goes beyond the standard, then a possible corrective action would be to change the policy and the auditor must accept this. The same argument applies to procedures that go beyond the standard. Some companies are content to have guidance policies but not mandatory policies and so may feel that there is lack of commitment if the non-product quality issues are removed from the mandatory procedures. An alternative nonconformity that does not require activities nonconforming with the procedure to be brought into line is to state that the procedures have not been maintained, since procedures do not match practice.

Procedures consistent with the standard

Documented procedures have to be prepared which are consistent with the requirements of the standard. This could be interpreted as implying that only where the standard requires documented procedures need documented procedures be prepared; or where a procedure is needed to implement a requirement, as in clause 4.2.3 where it requires the supplier to define and document how the requirements for quality will be met. This could imply that a procedure is required. The auditor cannot justify a nonconformity unless the standard specifically requires a documented procedure. The requirement can be met by a plan, a specification, a standard etc.

Content of procedures

In clauses 4.10.1, 4.13.1, 4.14.2 & 4.14.3 the standard requires procedures to include various things. In no other clause does the standard specify what must be included in procedures. There is an implication that where documented procedures are required they should respond to the requirements of the relevant clause but since it is not explicitly stated, there is no nonconformance if procedures do not respond to every requirement of the relevant clause. The requirements of the standard have to be addressed by the documented quality system whether in procedures or elsewhere. If procedures contain omissions but the requirements are addressed in the Quality Manual then unless it can be shown that the documented procedures are ineffective there is no nonconformity.

Implementing procedures

Suppliers have to implement effectively the quality system and its documented procedures, which could mean that some suppliers can limit their procedures to the requirements of the standard and others can go beyond the requirements of the standard, thereby creating inequality in quality system audits. The standard addresses provisions for achieving customer satisfaction. If the procedures go beyond measures to achieve customer satisfaction then these areas are of no concern to the auditor.

Documentary evidence of compliance

There is a requirement to implement effectively the quality system and its documented procedures but no requirement to maintain documentary evidence of compliance with the requirements of the standard unless the standard is one of the requirements specified in the contract or order. If documentary evidence is not available at the time of the audit, either because there has been no cause to perform the activities required by the system or because any evidence that did exist has not been retained, then there is no nonconformity. The only exception is where the standard requires specific quality records through a cross reference to clause 4.16 and where the stated retention time is more than the period between the creation of the records and the date of the audit.

Quality planning

Planning procedures

Documented procedures are not required for quality planning. The same argument that is used above for quality system development procedures applies.

Quality planning beyond the standard

If a supplier's quality system includes provisions beyond the standard then the quality planning requirements require that the quality planning be consistent with the quality system and therefore cannot be restricted to compliance with the standard. If the quality system addresses typical contract requirements that are beyond the standard then where quality planning does not address these requirements there is a nonconformity. If the system addresses requirements that do not serve customer satisfaction and are not addressed by quality planning then the auditor should write an observation rather than a nonconformity against the standard.

Considerations in quality planning

The standard requires that the supplier gives consideration to several activities in meeting specified requirements for products, projects or contracts but does not require that the result of giving such consideration be defined and documented. The auditor has to obtain reasoned argument from the auditee that consideration was given to the topics in clause 4.2.3 and for the reasons given, the course of action taken was decided.

Contract review

Procedures for contract amendments

Procedures have to be maintained for contract review and for co-ordinating these activities but not for making amendments to contract. The means by which amendments are made needs to be identified. This will usually be covered by the particular contract so any procedures needed will be specified by the customer.

Contract review procedures

Procedures have to be established for contract review and for the co-ordination of these activities but not to achieve any particular purpose, unlike the procedures that are required for design control and control of nonconforming product. The contract review procedures should cover the requirements of the standard and ensure that the only contracts that are accepted are those which the supplier has the capability of executing to the customer's satisfaction. If evidence exists to show that the procedure will not achieve this purpose, then it is nonconforming with the requirements of clause 4.3.2.

Meeting contract requirements

The standard requires the supplier to have the ability to meet contract requirements rather than meet specified requirements for the product, as indicated in clauses 4.2.3, 4.10.1, 4.10.4 etc. Where evidence is found that the supplier does not have the ability to meet a contract requirement, regardless of it affecting product quality, a nonconformity is justified against clause 4.3.2c.

Content of contract review records

The content of contract review records is not specified so can be an acceptance mark on a contract/order or a file of contract negotiations. The records need to demonstrate that the requirements of clause 4.3.2 were met as applicable, otherwise they do not conform to the definition of quality records in ISO 8402.

Amendment to a contract

Documenting contract amendments

There is no requirement for contract amendments to be documented. An amendment to a contract is a change to a document that relates to the requirements of the standard and therefore comes within the provisions of clause 4.5.3 on changes to documents.

Whilst the amendment itself does not have to be documented the change to the document does have to be approved prior to issue.

Approval of contract amendments

There is no requirement to review the amendment to contract before acceptance, only to review it prior to issue as stated in clause 4.5.2. An amendment to a contract creates a different contract and therefore it is required to be reviewed before acceptance.

Defining how contract amendments are made

Suppliers have to identify how an amendment to a contract is made but not define this in documented procedures. The same argument as given above applies.

Design control

Design verification procedures

The standard requires procedures for design control and verification implying that design control does not include verification. In principle design control does include design verification and validation since design verification and validation is the measurement component of the design process.

Design validation procedures

The design control element covers design verification and design validation but the procedures only have to cover design verification. This is an oversight since the design output has to be expressed in terms that can be verified and validated, so procedures should cover both.

Design and development planning

Documenting design and development plans

The standard requires the supplier to prepare plans for each design and development activity but does not require that these plans be documented and implemented. It could therefore be assumed that only if the design control procedures require the plan to be documented and implemented can a lack of its documentation and implementation be classed as a nonconformity with quality system procedures. As the system has to be documented, it follows that design and development plans have to be documented as they are part of the system and clause 4.2.2a requires the system to be implemented.

Duplication of assignment of personnel

Design and development activities are required to be assigned to qualified personnel equipped with adequate resources, which duplicates the combined requirements of clauses 4.1.2.2 and 4.18. If a nonconformity is detected on the assignment of qualified or trained personnel or the adequacy of resources the best clause should be chosen but the auditor should not prepare separate nonconformity statements against clauses 4.4.2, 4.1.2.2 and 4.18.

Organizational and technical interfaces

Organizational and technical interface information is required to be documented but the resources to implement the design and development plan do not have to be documented. Clause 4.4.2 requires the design and development activities to be assigned to qualified personnel equipped with adequate resources. Such resources should be identified to meet the requirements of clause 4.1.2.2; therefore if they are not documented their adequacy cannot be verified. Not only that, the management cannot be certain that they have correctly estimated the resources needed to design the product or service within the defined timescales. If such documents do not exists, there is no nonconformity but it has to be evident that the resources are adequate. The auditor has to exercise judgement and focus on aspects that may indicate a lack of resources, such as delays, backlogs and excessive overtime being worked. If the job cannot be done in normal working hours it usually indicates a resource problem.

Design input

Pre-release review

The design input requirements do not have to be reviewed for adequacy before implementation, as do design changes in clause 4.4.9 and design output documents in clause 4.4.5. Whilst not explicitly stated in this clause of the standard, since design input documents are those which relate to the requirements of the standard, the requirements of clause 4.5.2 do apply.

Inclusion of user needs

There is no requirement for design input requirements to include user needs and/or requirements, although the design output has to meet design input and must be expressed in terms that can be validated against user needs. This is an oversight and any approved and issued design input documentation that did not include user needs and/or requirements would be inadequate under clause 4.5.2.

Inclusion of design verification requirements

There is no requirement for design input requirements to include design verification requirements. Many design requirement specifications issued by customers include verification requirements since the assurance they require may need to be demonstrated in certain ways. Where verification requirements are absent, the supplier needs to have established that there are no customer verification requirements as part of the contract review process. This is all part of establishing that they have the capability of meeting the customer's requirements. Some customers may require a method of verification that cannot be performed by the supplier. Auditors therefore need to verify that design verification was considered during the contract review if verification requirements were omitted from the design input documents.

Design output

Design output documents are not required to be approved before release. Auditors should verify that design output documents are governed by the document and data control procedures, which should require both their review and approval before their release.

Design review

The content of design review records is not a specified requirement of the standard. As with the management review records, the design review records should represent the results of the review rather than a record of a meeting convened to discuss the results of the review. Auditors should examine the records against the ISO 8402 definition and, if inadequate, declare a nonconformity against this standard.

Design verification

The standard requires design verification measures to be recorded but it is unclear whether design verification requirements are to be specified or design verification results are to be recorded. As clause 4.16 is referenced the implication is that there is no requirement for design verification requirements to be defined and documented, since design verification is not addressed in the clause on design input. Clause 4.4.2 requires that the supplier prepares plans for each design and development activity. Design verification is one such activity; therefore the plan has to address design verification. However, the standard is weak in that it does not specify what in particular has to be addressed. If design verification requirements are not specified by the supplier, then the auditor has to decide whether the product characteristics and conditions of use are sufficiently detailed without additional verification requirements. If not then the nonconformity is against the adequacy of design control procedures in ensuring that specified requirements are met.

Design validation

Specification of user needs

The user needs and/or requirements are not required to be specified in the design input or the contract. This is dealt with under the heading *Inclusion of user needs*.

Documenting design validation requirements

There is no requirement for design validation requirements to be defined and documented. The lack of design validation requirements can be dealt with in the same way as for design verification requirements.

Recording design validation results

There is no requirement for design validation results to be recorded. This is an oversight. ISO 9001 is usually invoked in contracts for design, therefore the catch-all requirement in clause 4.16 for quality records covers this. However, if ISO 9001 is not invoked in the contact the auditor has to take the view that a design review should be performed after the validation stage and therefore design review records should be available which detail the results of the design validation activities.

Design changes

The title is 'design changes' whereas the text addresses 'design changes and modifications' and there is no definition in ISO 8402 of the difference between a design change and a modification. Auditors should take the view that design changes are changes to a design and modifications are changes to a product as a result of a design change. The supplier's design control procedures therefore need to cover both changes to designs and changes to products resulting from design changes.

Document and data control

External data

The requirements apply to external documents but not to external data. External data may be in the form of tables, technical information, statistics taken from external publications. These publications may change but the data may remain unchanged.

Since the governing requirement is that the supplier shall control all documents and data, the auditor should establish what external data is being used and verify that provisions have been made for capturing data changes.

Definition of data

The term 'data' and how it differs from a document is not defined in ISO 8402. Data is information organized in a form suitable for manual or computer analysis.

Document and data approval and issue

Use of documents and data

There is no requirement for documents and data to be reviewed and approved prior to use or release if they are not issued. The term 'issue' usually implies the distribution of information. If documents and data are being used prior to formal issue then the auditor needs to establish what they are being used for and make a judgement as to whether adequate controls exist to prevent the use of invalid or obsolete documents and data, whether or not it is reviewed, approved or issued.

Revision status of data

There is no requirement for the master list or document control procedure to identify the current revision status of data. Data often doesn't have any revision status but can be identified by its source and date. Where data is inserted into documents, the source should be stated. A list of data sources should therefore be maintained where such data is relevant to product/service quality. Data used in quality records does not need to be maintained since it is valid only at the time the record was produced. Data in specification, procedures, standards etc. used to generate product characteristics should be of known status.

Obsolete data

There is no requirement for the preclusion of obsolete data. Data can exist in documents but also in a computer database. Control of the database is important and therefore the general requirement for data control applies.

Availability of data

There is no requirement for pertinent issues of data to be available at all locations. The same argument as applied to obsolete data applies to the availability of data.

The Tale of the Missing Complaints

Andrea had worked in a Quality Department before becoming a third party auditor so felt confident in interviewing Richard, the Quality Manager. In fact, for him, the role of Quality Manager was not his primary job. He was also in charge of the sales department, which is why he felt comfortable when Andrea had suggested she examine the customer complaints system.

Andrea questioned Richard about the procedures employed for handling customer complaints. Richard reached out to grab a slim file and proceeded to explain the process. As Andrea listened she became suspicious that this file could not possibly be complete. Over a two-year period, the company had only received two complaints - one this year and one last.

Andrea asked to examine the letters of complaint and the forms the company had used to record the action they had taken. The action was clearly specified and there was an indication that its implementation had been checked. She asked what the criteria were for registering a complaint. Richard explained that a complaint is valid if sent to the MD in writing. Andrea then checked the procedure and confirmed this to be correct. In re-reading the forms Richard had presented, Andrea noticed that there was no action taken to prevent recurrence of the complaint. She asked Richard what the corrective action was and was directed to the statement on the form. She questioned Richard further and concluded that the action actually taken would not have prevented recurrence. Andrea noted these facts and moved on to her next assignment.

Meanwhile, Don, who had been out with the service engineers came back and he and Andrea compared notes as they prepared their report for the Closing Meeting. In talking through the events of the day, Don reported that the service engineers often received complaints from customers and made notes in their log books. He had noted down a few details and had intended to ask Andrea if these were logged in the customer complaints file. Andrea had felt all along that the customer complaint system was not effective and now she had the evidence she needed to prepare a nonconformity statement.

The moral of this story is:

Effectiveness is doing the right things not doing things right - you need to check both!

Removal or identification of obsolete data

There is no requirement for obsolete data to be promptly removed or otherwise identified. The same argument as applied to obsolete data applies to the removal of data.

Document and data changes

There is no requirement to identify the nature of change in data whether or not it is practicable. Data changes can be more difficult to detect if not identified. A new table of parameters may look the same as the old table. Only the decimal place may have been moved. Auditors should verify that the data control procedures cover this aspect. Where data changes are not identified, the data controls would not be adequate to cause the use of the changed data.

Purchasing

The standard does not specify whether the specified requirements with which the purchased product has to conform should be the customer, purchaser or supplier requirements or all three. Since the standard addresses a system that will ensure conforming product is supplied to customers, the specified requirements are those of the customer in this case. This will be made clear in the new version of ISO 9000-2.

Evaluation of subcontractors

Selection criteria

There is no requirement for the criteria for selecting subcontractors to be defined and documented. This may be included in the purchasing procedures but there is no requirement. Purchasing documents need to contain the subcontract requirements, any quality assurance requirements and quality system requirements as indicated in clause 4.6.3. Auditors should verify that the selection criteria are compatible with the criteria specified in the purchasing documents.

Documenting subcontractor controls

There is no requirement for the type and extent of control exercised by the supplier over subcontractors to be defined and documented. This too may be documented in the purchasing procedures but if not, the receipt inspection procedures should specify the acceptance criteria as required by clause 4.10.2.1 and the purchasing documents should specify any supplier verification on subcontractor's premises as required by clause 4.6.4.1.

Records of subcontractors

The requirement to establish and maintain quality records of acceptable subcontractors is ambiguous since there is no requirement to maintain records that demonstrate why the subcontractors are acceptable. The requirement is limited to quality records and does not extend to delivery records. A record of acceptable subcontractors should contain the objective evidence of their acceptability for the records to qualify as quality records as per ISO 8402. A list of names is insufficient without the acceptance criteria and performance record.

Purchasing data

Establishing purchasing documents

There is no requirement to establish and maintain purchasing documents, only a requirement for what they should contain. This is an oversight. However, many purchases are made over telephone lines without any documentation. Since information passed over telephone lines is data, it is therefore governed by the document and data control requirements of element 4.5.

Issue of purchasing documents

There is no requirement to issue purchasing documents to subcontractors. The phrase 'prior to release' can mean its release to subcontractors or its release for use, as in the case of a database that is used for purchasing purposes. There is no nonconformity if the data can be shown to have been reviewed and approved prior to use.

Supplier verification at subcontractors

There is no requirement for the supplier to plan and implement any proposed subcontractor verification, only a requirement to specify it. The subcontractor control plans form part of the documented quality system, although they are not procedures. Clause 4.2.2b requires that the quality system be implemented, so a failure to implement the plans would be a nonconformity with clause 4.2.2b.

Customer verification of subcontracted product

There is no requirement for the rights afforded to customers to verify subcontracted product to be specified in purchasing documents and conveyed to subcontractors. This is an option for the supplier and if such rights are not conveyed to the subcontractor, problems with customer access may occur. The auditor should establish how the supplier intends to deal with this.

Control of customer-supplied product

There is no requirement for records of customer supplied product to be maintained, only records of such product that is lost, damaged or is unsuitable. The contract should identify the products being supplied by the customer but if not, the auditor should verify that the inspection records under clause 4.10.5 include such details.

Product identification and traceability

Identifying product

There is no requirement to identify product only to establish and maintain procedures where appropriate, i.e. where the absence of procedures would adversely affect quality. This is an oversight however; unless the quality system provides for product to be identified, it will not cause the supply of conforming product and therefore is nonconforming with clause 4.2.1.

Relevance of product identification

The standard does not specify the circumstances where product identification is required. The auditor needs to establish whether the supplier's controls will prevent the supply of nonconforming product and if identity of the product is not obvious, the system is ineffective and nonconforming with clause 4.2.1.

Traceability requirement

The requirement for traceability is limited to where it is a specified requirement but the standard does not indicate whether the specified requirements in this case are those of the customer or those of the supplier. There are in some cases legal requirements for traceability: e.g. pharmaceuticals, components of transport vehicles. The auditor needs to establish the regulations that apply and verify that the system makes provision for compliance.

Process control

Incomplete scope

The heading to this element does not wholly reflect the scope, as the requirements do not refer to inspection as the method of measuring product characteristics. Inspection has been placed under a separate heading to avoid repetition under 'Purchasing and Process Control'. This is not intended to imply that inspection and test are not part of process control, but an explanation in this clause would have been useful.

Documented plans

There is no requirement for the supplier to document the plan of the production, installation and servicing processes. As the plan forms part of the quality system that has to be documented (clause 4.2.1) it follows that the plans for production, installation and servicing processes need to be documented.

Content of production, installation and servicing plans

There is no requirement for production, installation and servicing plans to define responsibility for their implementation, unlike the requirement on design and development plans in clause 4.4.2. Responsibilities for such tasks have to be documented as defined in clause 4.2.2.1, whether in these plans or in other documents. It is inconsistent, however, that design is treated differently. Should the auditor find that responsibilities are not defined in these plans then the supplier needs to demonstrate how the work specified will be carried out by the appropriate staff.

Qualified process staff

There is no requirement for production, installation and servicing activities to be assigned to qualified personnel equipped with adequate resources unlike the requirements for design and development activities of clause 4.4.2. Clause 4.18 requires personnel performing activities to be qualified only. The auditor needs to draw together the requirements of clauses 4.1.2.2 and 4.18 and verify that competent personnel equipped with adequate resources are assigned to carry out these activities.

Servicing procedures

Whereas this clause requires only documented procedures for servicing where their absence could adversely affect quality, clause 4.19 requires documented procedures regardless of such conditions. There are two types of procedures: control procedures and operating procedures. Clause 4.9 deals with the operating procedures and clause 4.19 the control procedures. However, there are no separate clauses on production and installation so there would appear to be another inconsistency here. The auditor needs to verify that the system has been documented as per clause 4.2.1 and if production and installation processes are not documented there is cause for a nonconformity.

Process monitoring records

There is no requirement to record the results of monitoring and controlling process and product parameters. Depending on the process, records may not be necessary since an on-line monitoring activity may cause the data to be refreshed continually. If statistical techniques are used then there have to be documented procedures but again there is no requirement to maintain records unless the process is classified as a special process. The absence of records does not indicate a lack of control providing the inspection and test records required by clause 4.10.5 are maintained. The auditor needs to establish whether a lack of records adversely affects the ability to control the process.

Records of equipment and environment

There is no requirement for records of the equipment used and the environmental conditions provided. The same argument as applied to process monitoring applies in this case. The auditor needs to establish whether a lack of records adversely affects the ability to control the process.

Records of approved processes and equipment

There is no requirement for records of approved processes and equipment. Without a record of some kind, the supplier will not be able to demonstrate that the processes and equipment in use have been approved. The process controls have to be documented by virtue of clause 4.2.1 and such documentation should include the process and equipment approval method. The 'as appropriate' in the requirement refers to the processes and equipment that need approval. Obviously simple, proprietary equipment and processes need only be selected, but their selection may need approval if there is a choice available.

Documenting workmanship criteria

There is no requirement for workmanship criteria to be defined and documented, only that they be stipulated. The term 'stipulated' means to lay down as a condition of an agreement. The agreement in this case is the procedure. Either the procedure needs to specify the criteria or should make reference to the model, illustration, sample etc. which is to be used. The absence of such identification indicates that the procedures are inadequate.

Records of equipment maintenance and process capability

There is no requirement for records of equipment maintenance and process capability to be maintained. The same argument as applied to process monitoring applies to this situation.

Determination of process capability

There is no requirement for process capability to be determined prior to commencement of production, installation or servicing, unlike the requirements in clause 4.11.1. This clause covers the use of measuring devices that have to be proven as capable of verifying the acceptability of product prior to release for production, installation or servicing. This is an oversight. The requirements for maintenance of equipment to ensure continuing process capability is premature if process capability had not been established in the first place. The auditor therefore has to establish that the process is not under adequate control since its capability was not established initially.

Documentation of process qualification

There is no requirement for the requirements for process qualification to be defined and documented, only that they be specified. The implication is that such process qualification be specified in the production plan or the documented procedures. If they are not then the auditor can identify a nonconformity with clause 4.2.1.

Receiving inspection and testing

There is no requirement for the recorded evidence of conformance obtained through subcontractor controls to be treated as a quality record as there is no reference to clause 4.16. Such evidence does constitute a quality record and thus is an oversight in the standard. If such recorded evidence is not maintained in accordance with clause 4.16, the auditor has to establish how the receiving inspection can be carried out without such data.

In-process inspection and testing

There is no requirement to perform in-process inspection, only to carry out any in-process inspection in accordance with documented procedures. In-process inspection is only necessary when product characteristics may not be acceptable for verification at subsequent stages. The auditor has to establish that where in-process inspections are not performed, all the product characteristics can in fact be verified on final inspection.

Final inspection and testing

Results of inspection and test

There is a requirement for the results of final inspection to meet specified requirements and not to demonstrate that the product meets specified requirements. The intention here is that the results are within the tolerances range specified for the particular parameters. The auditor needs to check inspection and test results to verify that none are outside specified limits. If any are found outside the limits then the supplier should have employed the nonconforming product controls to disposition the noncon-formity.

Dispatch criteria

The requirement for no product to be dispatched until all activities in the quality plan or documented procedures have been completed and associated data and documen-tation authorized is inconsistent with the inspection and test requirements. Such plans and procedures may specify activities that do constrain product dispatch. The auditor needs to establish what activities are required to be carried out and then judge whether any affecting the product remain incomplete. Requiring activities to be performed that do not add value is unprofessional.

Inspection and test records

Inspection authority

The cross reference to clause 4.16 implies that only the records that identify the authority for the release of product have to meet the requirements of clause 4.16 and not the other records referred to in this clause. This unfortunate cross reference is incorrectly placed and should have been on the first line of this clause. 'Release' in this case means release to the next stage in the process rather than release to the customer. Auditors faced with a supplier that interprets this differently may have more problems to deal with and should consider whether the supplier is really serious about quality.

Identity of measuring equipment

There is no requirement for details of the measuring devices used to carry out the inspections and tests to be recorded although, in some cases, it may be impossible to comply with clause 4.11.2f if this were not done. Auditors should establish what the supplier's policy on this point is. It may be that there are few measuring devices so that such traceability is unnecessary. Where there are many such devices, traceability is much more important.

Control of inspection measuring and test equipment

Scope of requirement

This requirement addresses more than the measurement devices referred to as it contains requirements for measurements to be made which may be defined during design. Auditors should be aware that suppliers may misinterpret these requirements.

Defining the calibration process

There is no requirement to document the calibration process, only to define it As with other examples clause 4.2.1 requires the system to be documented, so the calibration process needs to be documented as well as defined.

Inspection and test status

There is an implication that product that does not meet requirements can be released under an authorized concession but there is no requirement in clause 4.13.2 defining who should authorize the concession. The implication is that authorization by the customer is required, but it can be anyone who has been authorized through meeting the requirements of clause 4.13.2. In some cases, suppliers may authorize concessions without referral to their customer. This is valid where the customer has not imposed any specific requirements and is purchasing product to the supplier's specification. The auditor needs to verify that such concessions do not conflict with the description of the product as advertised, otherwise there is a nonconformity.

Control of nonconforming product

Responsibility and qualification for disposition action

The requirement for the responsibility for review and authority for the disposition of nonconforming product to be defined duplicates that given in clause 4.1.2.1. The requirement should have indicated that those performing the review and making the decision need to be qualified. The auditor should verify not only that such personnel are authorized but are also suitably qualified.

Disposition procedures

The requirement for the documented procedures to provide for the disposition of nonconforming product is incomplete, as in clause 4.13.2 the review of nonconforming product and the inspection of repaired and/or reworked product is also required to be in accordance with documented procedures. The supplier's procedures for the control of nonconforming product should cover the disposition process, including the decisions, the execution of the decisions and subsequent verification.

Specified requirements

The context in which the term 'specified requirements' is used in clause 4.13.2 implies that they are customer requirements and not supplier requirements, otherwise there is no reason to report the nonconformance to the customer for concession. A supplier could argue that the nonconforming product controls only apply to deviations from customer requirements and they would be correct. However, in many cases the achievement of customer requirements occurs throughout the process, not just at the end, and in such circumstances element 4.13 applies throughout the process.

Acceptance of concessions

There is no requirement for the concession reported to the customer to be accepted by the customer. Clause 4.12 permits nonconforming product to be released under an authorized concession but does not specify that the customer should have authorized it. The contract will usually define the acceptance criteria and the arrangements for delivery of nonconforming product. Where there are no such contract conditions then the customer does not have to be approached. (See also previous comment under *Inspection and test status*.)

Corrective and preventive action

Documenting corrective and preventive actions

There is no requirement for the corrective and preventive action procedures to require proposed corrective and preventive actions to be defined and documented. This is an oversight. However, an action may be difficult to implement reliably and timely if not documented. The auditor needs to examine the procedures and establish whether, in the circumstances, the effectiveness of the quality system is impaired by not documenting the corrective and preventive actions to be taken.

Records of corrective and preventive action

There is no requirement to record the corrective and preventive action taken, only those which result in changing documented procedures. Again an oversight. Records will usually exist but as there is no reference to clause 4.16, they do not have to be maintained for a defined period. The auditor should sample a few investigation records and verify that the nonconformity has been eliminated. In the case of preventive actions, the auditor should sample the management review records to see whether the preventive actions have been recorded. The auditor should establish whether a failure to document corrective actions impairs the effectiveness of the system and if so it warrants a nonconformity under clause 4.1.3.

Corrective action

Reports of product nonconformities

The requirement for the corrective action procedure to include the effective handling of reports of product nonconformities duplicates the requirement in clause 4.13.1 for documented procedures for nonconforming product unless the requirement of clause 4.14.2 is intended to apply only after delivery of product to customers. Auditors should assume that this requirement applies to external reports of product nonconformities and clause 4.13 applies to internal reports of nonconformities.

Records of corrective action

There is no requirement to record the actual corrective action taken, unlike in clause 4.13.2 which requires the actual condition of accepted product to be recorded. The same argument as used above applies.

Submission of corrective action information

There is no requirement for the relevant information on corrective actions taken to be submitted for management review, unlike the requirements of clause 4.14.3d for preventive action. This is an oversight. However, since the management review is required to be of the quality system, then all aspects of the system should be reviewed, not just those where indicated in the standard.

Initiation of corrective actions

There is no requirement for corrective actions to be initiated, unlike clause 4.14.3c for preventive actions: an oversight but not a serious one. Control cannot be applied to something that has not been initiated.

Preventive action

The requirement to determine the steps needed to deal with any problems requiring preventive action extends the scope of the preventive action procedures beyond potential nonconformities and it does not limit it to quality problems. The auditor should refer to ISO 8402 for a definition of preventive action before requiring suppliers to deal with all problems whatever their source. Suppliers can only take action on matters within their control.

Handling, storage, packaging, preservation and delivery

Inadequate storage areas

The standard does not require the use of designated storage areas or stock rooms to prevent loss of product, unlike clause 4.16 which requires quality records to be stored in facilities that prevent loss. The auditor should establish what the stores are used for and, if containing finished product or customer supplied product, then provision needs to be made to prevent loss; otherwise, contract deliveries will be jeopardized and clause 4.3.2 requirements will not be met.

Conformance to specified packaging requirements

Packing, packaging and marking processes are required to be controlled only to ensure conformance to specified requirements and not to provide protection to and identity of the packaged product. The auditor should establish what the contract or the market requires and verify that the packaging and marking is compatible. The purpose of packaging is to protect the product so if it appears inadequate, even though meeting the specified requirements, there is a nonconformity with clause 4.15.6.

Under the supplier's control

Preservation and segregation is required only when the product is under the supplier's control but this is true also of storage, handling and packaging. This is an oversight and should have appeared in clause 4.15.1.

Applicability of handling, storage and packaging requirements

The requirement for protection of the quality of the product after final inspection and test duplicates the requirements for handling, storage, packaging and preservation. These requirements are not limited to protection of product after final inspection and test. The auditor should verify that the supplier has applied these requirements from receipt of product up to the stage when their responsibility ceases.

Control of quality records

Records to demonstrate conformance

The requirement for quality records to be maintained to demonstrate conformance to specified requirements conflicts with the cross references to clause 4.16 within the standard. All the cross references relate to where records are required to be maintained. This requirement, however, implies that whatever the specified requirement, quality records have to be maintained to demonstrate conformance, even with aspects that do not affect product or service quality if specified in a contract, order or documented procedure. Auditors should be aware that ISO 9001 addresses quality assurance requirements and not quality management requirements. The requirements if met are intended to give assurance to customers that products supplied *will* meet their requirements. If conformance with a requirement cannot be demonstrated by the supplier it does not follow that the supplier is noncompliant. The requirement may not apply or its implementation may not yet be accomplished. Even where implementation has been accomplished, if there is no requirement for a record then the auditor cannot insist there should be one. The auditor has to establish that operations are under control and this can be done without there being records (see Chapter 4 under *Objective evidence*).

Subcontractor records

The requirement for pertinent quality records from the subcontractor to be an element of the supplier's quality records applies only where such records have been supplied; therefore the term 'where appropriate' should have been used. Suppliers cannot be expected to include records that they do not possess. Typical records that may well be supplied are certificates of conformity and inspection and test results but only if required to do so by the supplier.

Loss of quality records

The requirements for the storage of quality records are more onerous than for the storage of product, since such areas are required to prevent loss. Whilst this may seem at odds with the spirit of the standard, products can be replaced, whereas records cannot. Records are results of an event that has taken place. Unless one can repeat the event, lost records are lost for good.

Internal quality audits

Planned arrangements

The requirement for quality activities and related results to comply with planned arrangements is inconsistent with other requirements of the standard which requires activities to carried out in accordance with documented procedures and results to comply with specified requirements. The term 'planned arrangements' is shorthand for policies, procedures, specifications, contracts, instructions etc. The term 'specified requirements' could have been used but this term seems to be used only when related to the product being supplied and not activities.

Audit schedule

There is no requirement for the audit schedule to be defined and documented. It has been stated several times that the quality system has to be documented and, as the audit schedule is part of the system, it too must be documented as required by clause 4.2.1.

Documenting corrective actions

There is no requirement for corrective actions resulting from deficiencies found during the audit to be defined and documented. This omission is similar to that in clause 4.14.2. Auditors can obtain evidence that corrective actions have achieved their objectives by examination of the follow-up audit records. If the failure to document corrective actions impairs the effectiveness of the system a nonconformity could be stated against clause 4.1.3 or the finding could be treated as an observation.

Training

Training procedures

There is no requirement for documented procedures for the provision and control of training and for evaluating its effectiveness, only for identifying training needs. If the auditor can show that the effectiveness of the system is impaired by a lack of training procedures then a nonconformity could be stated against clause 4.2.1 to the effect that the system for training has not been documented. Alternatively, clause 4.2.2 does require procedures to be dependent upon complexity and methods used, so if the auditor has evidence that the training of staff is complex or uses methods that would not be implemented effectively unless documented in a procedure, then a nonconformity could be stated against this clause.

Documenting personnel qualifications

There is no requirement for suppliers to define and document the qualifications required of personnel performing specific assigned tasks. The supplier needs to specify the qualifications required for staff to perform their assigned tasks. Whether this is done through verbal means or through a job specification depends upon the complexity of the system. The system has to be documented so that a policy stating how the qualifications for various jobs are established may suffice. The auditor cannot insist on there being job specifications.

Appropriate training records

If suppliers believe it is not appropriate to maintain training records for certain types of training then the standard gives them this right. Appropriate records of training does not mean appropriate training records. Appropriate records of training means that records of training have to be maintained but their content and format should be appropriate to the circumstances. In some cases, these records may be no more than a list of personnel undertaking a training course. In other cases it may need to contain levels of competency. The auditor should establish what is appropriate in the circumstances and avoid being over prescriptive. The records need only record training that is relevant to the job the personnel are performing. Since personnel can be qualified on the basis of education, training and experience then in some cases there may be no training records for an individual. There is no noncompliance. Training records can only exist after a person has been trained (see also under *Resources*).

Servicing

There is no requirement for records to be maintained of servicing activities, only reports that servicing meets specified requirements. Servicing should be treated as a group of processes that are governed by the requirements of clause 4.9. Although there are few requirements in clause 4.9 for quality records, the associated inspection, test, nonconforming product controls etc. should produce sufficient records to demonstrate that servicing operations are under control.

Statistical techniques

There is no requirement for documented procedures that enable the need for statistical techniques to be identified or for such needs to be documented. This inconsistency is the opposite to that on training where procedures are required for identifying needs. It would be logical for the need for statistical techniques to be identified within the relevant control procedures – such as design, purchasing, production, installation and servicing procedures – but other means are equally valid. The auditor should verify that provisions are included in quality system documentation for such needs to be identified.

Chapter 9

Check lists for auditors

Introduction

Check lists serve as an aid to memory and may be generic or custom-made for specific situations. A combination of both is the norm. The preparation of check lists was dealt with in Chapter 3. This chapter contains a series of generic check lists based on the requirements of ISO 9001.

Each check list refers to one of the process groups identified in the business management system model given in Chapter 3. The lists contain topics to be addressed rather than questions to be asked. In using these lists, auditors will therefore have to derive questions on a particular topic to explore the subject. As general rule, the topic can be preceded with phrases such as: 'How do you ...?'; 'What are your ...?'; 'Who is responsible for ...?'; etc. Guidance is given on their application and explanatory notes added where appropriate. Further guidance, hints and tips are contained in Chapter 10.

These lists should therefore be applicable to any type of business although tailoring will be needed for specific situations. None of the check lists includes specific topics concerning supplier procedures, policies, technology etc. These will need to be added to ensure adequate coverage for a specific audit.

It will be noted that each list contains identical or similar topics in addition to the process-related topics. This arises due to several of the requirements of the standard being applicable to all processes. Auditors may choose to ignore common topics in some processes providing a representative sample is obtained across the organization.

Providing probing questions are asked against each topic, the checks should reveal the objective evidence needed to verify conformity with the identified clauses of ISO 9001.

Business management

Application guidance

Business management is the name given to that area of the business that provides the direction, policy and strategy. Often the Lead Auditor will deal with this aspect of the audit. Whilst ISO 9001 does not address business management specifically, the requirements relating to quality policy, responsibilities, resources etc. need to be verified within the area of top management. This check list therefore aims to aid the examination of this important part of any business. The topics are listed in the order in which it is believed they should be addressed. Starting an interview with top management by asking them about the quality policy may in fact drive the interview in the wrong direction. Starting by asking them about their customers and whether they have a method of determining their customer's expectations, can lead into questions on the quality policy.

The purpose of these checks is to establish that top management are committed to the pursuit of quality and are directing its achievement, control, assurance and improvement.

Notes on the check list

Some of the entries in the check list may appear to be at odds with the corresponding requirements of the standard. Many of the checks serve to explore the methods of management so that consistency can be verified against outputs from other checks.

The auditor needs to establish customer expectations so that the relevance of the quality policy and objectives can be established.

The products of management are decisions but the standard does not refer to there being any control required over decisions, only documents and data. To understand whether a supplier is paying lip service to quality policy and the quality system or whether it forms an integral part of the business the auditor needs to understand how decisions are made, promulgated and changed in the company. Having explored that process he/she can then proceed to examine the quality policy.

No documentation is required to meet the resource requirements but to test whether the resources are adequate during the interview with the management, the auditor should establish whether the system provides for feedback concerning inadequate resources to be conveyed to management for action.

Item	Factor	Topic	Clause
1	Product	Strategic decisions concerning products/services,	4.1.1
		their marketing and distribution	4.1.2.2
2	Process	Determining customer expectations and needs	4.1.1
3		Capturing governing legislation, regulations, statutes etc.	4.1.1
			4.4.4
4		Determining organizational goals	4.1.1
5		Involvement of personnel in decision making	4.1.1
6		Promulgation of decisions	4.1.1
7		Forming quality policy	4.1.1
8		Promulgating quality policy	4.1.1
9		Changing decisions to maintain policy	4.1.1
10		Deviations from policy	4.1.1
11		Quality objectives	4.1.1
12		Quality system – relationship with other systems	4.2.1
13	Resources	Identifying resource requirements	4.1.2.2
14		Resource acquisition	4.1.2.2
15		Assigning trained personnel to management positions	4.1.2.2
16		Providing training resource	4.18
17	Measurement	Audit of decisions on use of authority	4.17
18		Audit of organization structures	4.17
19		Feedback on resource limitations	4.1.2.2
20		Problem identification and reporting	4.1.2.1
21		Dealing with customer complaints	4.14.2
22		Management review process	4.1.3
23		Management review intervals	4.1.3
24		Management review participants	4.1.3
25		Achievement of objectives	4.1.3
26		Determination of system effectiveness	4.1.3
27		Customer satisfaction measures	4.1.3
28		Corrective action	4.14.2
29		Preventive action	4.14.3
30	Personnel	Organization structure (inter-relationships)	4.1.2.1
31		Assigning responsibility	4.1.2.1
32		Delegation of authority	4.1.2.1
33		Appointment of Management Representative	4.1.2.3
34		Training policy	4.18
35		Identifying training needs	4.18
36		Commitment of management and staff	4.1.1
37	Documents	Availability of management documents	4.5.1
38		Procedure for controlling management documents	4.5.1
39		Quality policy and objectives	4.1.1
40		Responsibilities and authority documents	4.1.2.1
41		Organization charts etc. showing interrelationships	4.1.2.1
42		Issue of the management documents	4.5.2
43		Changes to management documents	4.5.3
44		Revision control	4.5.2
45		External documents used in business management	4.5.1
46		Obsolete documents	4.5.2
47	Data	Sources of data used in decision making	4.5.1
48		Electronic storage of data	4.5.1

Table 9-1 Business management check list (continues)

Item	Factor	Topic	Clause
49		Control of data	4.5.2
50		Changes to data	4.5.3
51	Records	Management review records	4.1.3
52		Training records for top management	4.18
53		Audit reports of organization audits	4.17
54		Investigation of product nonconformities reported by customers	4.14.2b

Table 9-1   Business management check list (continued)

Quality system management

Application guidance

Quality system management is the name given to that area of the business that manages the quality system. The Lead Auditor will often cover this aspect through an interview with the Management Representative and the internal auditors. The interview with top management should have established where their priorities lie with respect to quality and whether the quality system is truly integrated into the business. The role of the auditor in this interview is to obtain confidence that this system is under control and that it is not just the brain-child of one person. The order in which the topics are addressed in this case can be varied but should commence with an examination of the documented quality system.

The purpose of this check list is to establish that the quality system has been planned, organized, resourced and is being controlled in a manner that will enable it to fulfil its purpose.

Notes on the check list

The products of quality system management are the Quality Manual, procedures and records.

Documentation in this case refers to the control exercised over these documents and, where the system documentation is stored electronically on disk, then the auditor has to probe the integrity of the data controls.

System changes are far wider in nature than document changes since the impact of change on the organization, responsibilities, processes and resources needs to be examined.

Item	Factor	Topic	Clause
1	Product	Constituents of the documented quality system	4.2.1
2		Purpose and scope of the quality system	4.2.1
3	Process	Establishing the system - methodology	4.2.1
4		Planning system development	4.2.1
5		Relationship to quality policy and objectives	4.2.2
6		Influence of customer expectations	4.1.1
7		Establishing specified requirements	4.2.1
8		Documentation structure	4.2.1
9		Quality Manual coverage	4.2.1
10		Identifying need for procedures	4.2.2
11		Procedures references in manual	4.2.1
12		System changes	4.2.1
13		Quality records collection	4.16
14		Quality record storage	4.16
15		Availability of quality records to customers	4.16
16		Retention periods for quality records	4.16
17		Disposal of quality records	4.16
18	Resources	Identifying resource requirements for design, implementation and audit of quality system	4.1.2.2
19		Assigning trained personnel	4.1.2.2
20		Adequacy of staffing levels	4.1.2.2
21	Measurement	Internal audit schedules	4.17
22		Scope of internal audits	4.17
23		Audit planning	4.17
24		Independence of auditors	4.17
25		Implementation of audits	4.17
26		Audit results	4.17
27		Determination of system effectiveness	4.17
28		Timely corrective action for internal audits	4.17
29		Corrective action for external audits	4.14.2
30		Effectiveness of corrective actions from audits	4.17
31		Effectiveness of corrective actions controls	4.14.2d
32		Follow-up audits	4.17
33		Data input to quality system review	4.1.3
34		Analysis of system performance data	4.1.3
35		Reporting system performance to management	4.1.2.3 4.14.3d
36		Analysis of potential nonconformities	4.14.3a
37		Preventive action programmes	4.14.3b
38		Effectiveness of preventive action controls	4.14.c
39		Improvements to the quality system	4.1.2.3
40	Personnel	Authority of Management Representative	4.1.2.3
41		Qualifications of Management Representative	4.18
42		Authority of auditors	4.1.2.1
43		Qualifications of auditors	4.18
44		Authority to identify system problems	4.1.2.1
45		Authority to initiate preventive action on the system	4.1.2.1
46		Authority to recommend solutions to system problems	4.1.2.1
47		Authority to verify solutions to system problems	4.1.2.1

Table 9-2 Quality system management check list (continues)

Item	Factor	Topic	Clause
48		Identifying training needs for system development and implementation	4.18
49		Understanding quality policy	4.1.1
50	Documents	Availability of documents	4.5.2
51		Internal audit procedure	4.17
52		Corrective action procedures	4.14.1
53		Preventive action procedures	4.14.1
54		Procedure for the control of quality system documents	4.5.1
55		Issue of Quality Manual and procedures	4.5.2
56		Changes to Quality Manual and procedures	4.5.3
57		Revision control	4.5.2
58		Obsolete documents	4.5.2
59		External documents used in quality system management	4.5.1
60	Data	Sources of data used in quality system management	4.5.1
61		Electronic storage of data	4.5.1
62		Control of data	4.5.2
63		Changes to data	4.5.3
64	Records	Audit reports for external audits	4.17
65		Follow-up audit reports	4.17
66		Training records for auditors and MR	4.18

Table 9-2 Quality system management check list (continued)

Marketing

Application guidance

Marketing is the name given to that area of the business that attracts customers to purchase its products and services. There are few requirements in the standard that relate to marketing so some creative thinking is needed. The marketing function is usually responsible for determining customer expectations; therefore their output must be an input to the determination of the quality policy and objectives. Often customer expectations are set by what a company offers and so any data issued by the company that serves this purpose is relevant to the quality system. Tendering precedes the award of a contract. It is often not a feature of many organizations, but even those that supply proprietary products and services sometimes do receive invitations to tender for a contract. The activities of the sales personnel before a contract or order is received are also relevant in ensuring the organization has the capability to meet the orders secured.

Whilst the Lead Auditor will have established the method by which customer expectations are determined, the auditor assigned to interview marketing personnel needs to establish that there is a system in place for performing this task.

The purpose of this check list is to establish that the manner in which customer expectations are set and quality characteristics established is under control.

Notes on check list

The result should be a contract or order but this is covered in the Order Processing Check List. System effectiveness is the sum of the effectiveness of the component parts; therefore the effectiveness of the marketing practices needs to be determined.

Item	Factor	Topic	Clause
1	Product	Types of contracts and orders pursued	
2	Process	Establishing customer expectations and needs	4.1.1
3		Publicity material - capability claims	4.3.2c
4		Identifying product/service quality characteristics	4.3.2a
5		Receipt of invitations to tender	4.3.2
6		Preparation of tender documents	4.3.2
7		Review of tender requirements	4.3.2
8		Consultation in review process	4.3.1
9		Determination of capability	4.3.2c
10		Promulgation of capability data	4.3.2c
11		Documenting customer requirements	4.3.2a
12	Resources	Identifying sales and marketing resources	4.1.2.2
13		Assignment of trained personnel	4.1.2.2
14		Adequacy of staffing levels	4.1.2.2
15		Identifying resource requirements to execute potential orders	4.1.2.2
16	Measurement	Audit of sales and marketing operations	4.17
17		Customer complaints	4.14.2a
18		Reports of product nonconformities	4.14.2a
19		Corrective action	4.14.2
20		Preventive action	4.14.3
21		Determination of marketing system effectiveness	4.1.3
22	Personnel	Defined responsibility and authority for marketing personnel	4.1.2.1
23		Authority to bid for a contract	4.1.2.1
24		Authority to approve tender documentation	4.1.2.1
25		Understanding quality policy	4.1.1
26		Authority of sales personnel	4.1.2.1
27		Qualifications of marketing and sales personnel	4.18
28		Identifying training needs	4.18
29		Authority to identify marketing problems	4.1.2.1
30		Authority to initiate preventive actions	4.1.2.1
31		Authority to recommend solutions to problems	4.1.2.1
32		Authority to verify solutions to problems	4.1.2.1
33	Documents	Availability of documents	4.5.2
34		Control of publicity material	4.5.1
35		Tender review procedures	4.3.1

Table 9-3 Marketing check list (continues)

Item	Factor	Topic	Clause
36		Tender requirement documents	4.3.2
37		Procedure for the control of tender documentation	4.5.1
38		Issue of tender documents	4.5.2
39		Changes to tender documentation	4.5.3
40		Revision control	4.5.2
41		Obsolete documents	4.5.2
42		External documents used in sales and marketing	4.5.1
43	Data	Sources of data used in sales and marketing	4.5.1
44		Electronic storage of data	4.5.1
45		Control of data	4.5.2
46		Changes to data	4.5.3
47	Equipment	Validity of exhibition materials and equipment	4.3.2
48	Records	Tender review	4.3.2
49		Training records for sales and marketing personnel	4.18
50		Audit reports of marketing audits	4.17

Table 9-3 Marketing check list (continued)

Order processing

Application guidance

Order processing is the name given to that area of the business that receives customer orders or contracts. It follows the marketing process but may well be part of it, depending on the way the organization is structured. Many of the same topics need to be addressed in order processing as were addressed in marketing but the emphasis is different. Once the order has been placed the company then enters into a commitment to execute it.

The purpose of this check list is to establish that adequate provision has been made to prevent the organization entering into contracts it is unable to fulfil.

Notes on check list

This check list is very similar to that for Marketing. However, it is important to separate the two processes where tendering is carried out by the organization. The impact of errors subsequent to contract award is far greater than in the tendering phase. Auditors should be able to verify that the company has the capability of meeting its customer requirements. It is therefore important for the auditor to examine particular contracts and establish the type of work being undertaken and the corresponding requirements before auditing other processes.

Item	Factor	Topic	Clause
1	Product	Types of contracts/orders processed	
2	Process	Establishing customer expectations and needs	4.1.1
3		Publicity material - capability claims	4.3.2c
4		Identifying product/service quality characteristics	4.3.2a
5		Receipt of contracts/orders	4.3.2
6		Review of contract/order requirements	4.3.2
7		Consultation in review process (co-ordination)	4.3.1
8		Determination of capability	4.3.2c
9		Promulgation of capability data	4.3.2c
10		Documenting customer requirements	4.3.2a
11		Agreeing verbal requirements	4.3.2a
12	Resources	Identifying order processing resources	4.1.2.2
13		Assignment of trained personnel	4.1.2.2
14		Adequacy of staffing levels	4.1.2.2
15		Identifying resource requirements to execute accepted orders	4.1.2.2
16	Measurement	Audit of order processing operations	4.17
17		Corrective action	4.14.2
18		Preventive action	4.14.3
19		Determination of order processing system effectiveness	4.1.3
20	Personnel	Defined responsibility and authority for order processing personnel	4.1.2.1
21		Authority to accept a contract/order	4.1.2.1
22		Authority to approve contract documentation	4.1.2.1
23		Understanding quality policy	4.1.1
24		Qualifications of order processing personnel	4.18
25		Identifying training needs	4.18
26		Authority to identify order processing problems	4.1.2.1
27		Authority to initiate preventive actions	4.1.2.1
28		Authority to recommend solutions to contractual problems	4.1.2.1
29		Authority to verify solutions to contractual problems	4.1.2.1
30	Documents	Availability of documents	4.5.2
31		Control of publicity material	4.5.1
32		Contract/order review procedures	4.3.1
33		Procedure for the control of contract documents	4.5.1
34		Issue of contract/order documents	4.5.2
35		Method of amending contract/order	4.3.3
36		Changes to contract/order documentation	4.5.3 4.3.3
37		Revision control	4.5.2
38		Issue of contract/order changes	4.3.3
39		Obsolete documents	4.5.2
40		External documents used in order processing	4.5.1
41	Data	Sources of data used in order processing	4.5.1
42		Electronic storage of data	4.5.1
43		Control of data	4.5.2
44		Changes to data	4.5.3
45	Records	Contract review	4.3.4
46		Training records for order processing staff	4.18
47		Audit reports for order processing audits	4.17

Table 9-4 Order processing check list

Order planning

Application guidance

Order planning is the name given to that area of the business that makes provision for meeting the requirements of particular orders. There may be one planning process or several depending upon the complexity of the work and type of contract or order. Order planning may be subdivided into:

- Quality planning. This is covered by clause 4.2.3 of ISO 9001.

- Design planning. This is covered by clause 4.4.2 of ISO 9001 but also clause 4.2.3.

- Production planning, installation planning and servicing planning. These aspects are covered by clause 4.9 of ISO 9001, but much of 4.2.3 also deals with these issues.

In some cases planning is carried out for each order and in others it is carried out against a forecast of orders for existing products and services. The check list can therefore be used in several processes and is therefore more of a generic check list than a specific one.

The purpose of this check list is to establish that adequate provision has been made to execute contracts before work commences.

Notes on check list

The product of the order planning process is the plan or series of plans and therefore they should exhibit all the necessary characteristics to ensure that adequate provisions have been made for meeting the customer requirements.

Item	Factor	Topic	Clause
1	Product	Types of plans for achieving contract requirements	4.2.3a
			4.4.2
2	Process	Method of meeting requirements for quality	4.2.3
3		Consistency of planning with quality system	4.2.3
4		Identification of new controls including procedures	4.2.3b
5		Compatibility of processes with contract requirements	4.2.3c
6		Development of new techniques, instrumentation	4.2.3d
7		Identification of new measurement capability	4.2.3e
8		Verification requirements	4.2.3f
			4.10.1
			4.10.4
9		Acceptance standards	4.2.3g
10		Identifying new quality records	4.2.3h
11	Resources	Identifying resources needed to implement plans including training needs	4.1.2.2
			4.2.3b
			4.18
12		Assignment of trained and qualified personnel	4.1.2.2
			4.4.2
13		Adequacy of staffing levels	4.1.2.2
			4.4.2
14		Adequacy of material, equipment and other resources	4.4.2
15	Measurement	Audit of order planning operations	4.17
16		Corrective action	4.14.2
17		Preventive action	4.14.3
18		Determination of order planning system effectiveness	4.1.3
19	Personnel	Defined responsibility and authority for order planning personnel	4.1.2.1
20		Understanding quality policy	4.1.1
21		Qualifications of order planning personnel	4.18
22		Identifying training needs of planning staff	4.18
23		Authority to identify order planning problems	4.1.2.1
24		Authority to initiate preventive actions	4.1.2.1
25		Authority to recommend solutions to planning problems	4.1.2.1
26		Authority to verify solutions to planning problems	4.1.2.1
27	Documents	Availability of documents	4.5.2
28		Procedures for the control of planning documents	4.5.1
29		Issue of plans	4.5.2
30		Changes to plans	4.5.3
31		Revision control	4.5.2
32		Updating plans (design only)	4.4.2
33		Control procedure for planning (design only)	4.4.2
34		External documents used in order planning	4.5.1
35		Obsolete documents	4.5.2
36	Data	Sources of data used in order planning	4.5.1
37		Electronic storage of data	4.5.1
38		Control of data	4.5.2
39		Changes to data	4.5.3
40	Records	Training records of planning staff	4.18
41		Audit reports for planning processes	4.17
42		Subcontractor records of planning activities	4.16

Table 9-5 Order planning check list

The Tale of the Hot Closing Meeting

It was a very hot day. So hot in fact that Tony and his team deliberately stayed in the air-conditioned areas to avoid being exposed to the sun. They had covered all the areas they had planned to cover and were looking forward to the Closing Meeting and going home.

At the review meeting Tony's team were busy compiling their reports and when they were completed, he called in the Quality Manager. In this small room, the heat was just bearable. The air-conditioning was a little antiquated but it worked. In 15 minutes Tony had agreed all the findings with the Quality Manager and they prepared for the Closing Meeting with the Managing Director. This was the first assessment of this company so all the staff were very keen to hear the results. Tony, however, was not prepared for what followed.

The MD walked in, followed by his immediate managers, the guides and some of the supervisors. When they had settled in their seats, Tony began his report. He had only reached item two on his agenda when the door opened and more staff came in. By now the air was very warm so the air-conditioning was turned on higher.

Tony had just started reporting the nonconformities when further people came into the room. There weren't enough chairs so more were brought in. The air-conditioning was by now hopelessly inadequate and it was already very noisy. Tony and his team were now squashed around the long teak table in the centre.

Tony asked someone close by if the room could be made cooler and within 5 minutes four pedestal fans were brought in. Tony felt that this would improve matters, but as soon as they were switched on he realized his mistake. They added to the noise and blew his papers across the room. He managed as best he could by indicating to another member of his team to take over. With all the noise, Tony could not hear what his fellow auditor was saying. The MD had a smile on his face so was obviously delighted with the report. They had intended to report that the company would not be recommended for registration but with the heat and the noise, Tony was uncertain whether this message was actually communicated.

The team quickly finished their report and were glad to leave the company.

The moral of this story is:

If there is bad news to deliver, make sure it can be conveyed effectively and gracefully!

Product/service generation

Application guidance

Product and service generation is the name given to that area of the business that converts plans into products and services. As the design process is fundamentally different from the production process, two separate check lists are needed. Design is a process of exploration, creating new products or services out of an idea or a requirement. It is a process that creates new wants, new standards and as such can be a journey into the unknown. Production is or should be a routine process which replicates an existing design that has been proven to work. It is a process that is cycled repeatedly and each time exactly the same. The design process will be recycled but in an iterative fashion, seeking improvement at every cycle. Production processes follow a proven path with a predictable outcome. In design the outcome is always uncertain. Another reason is simply that the requirements of the standard are different for design operations.

Installation is a form of production process but may be only cycled once, as will be the case with a large project such as a bridge, a power station etc. Other installations may be routine, such as installing software on a PC, and others may be a combination of the two, such as an electrical installation in a house where the components are standard but the layout different each time.

Servicing can be a routine process such as motor vehicle servicing or can be a one-off with unknown problems such as the servicing of a railway tunnel.

The design check list should be used for creative activities and the production check list for repetitive activities.

If the design is the output of the organization then in addition the order fulfilment check list should be used.

The purpose of these check lists is to establish that product/service generation processes are under control and will result in outputs that meet specified requirements.

Notes on check list

Inspection and test operations are classed as part of the process rather than as the measurement component but it matters not the category in which they are placed providing they are audited as an integral part of process control.

Item	Factor	Topic	Clause
1	Product	Types of product design, service design, packaging design	4.4.1
			4.15.4
2	Process	Design input	4.4.4
3		Consideration of contract review outputs	4.4.4
4		Statutory and regulatory requirements	4.4.4
5		Design controls	4.4.1
6		Design standards of acceptability	4.2.3g
7		Design and development planning	4.4.2
8		Control of subcontracted design	4.6.2
9		Control of technical interfaces	4.4.3
10		Product/service identification	4.8
11		Design output	4.4.5
12		Critical characteristics	4.4.5
13		Design changes	4.4.9
14	Resources	Identifying resources needed to design products and services	4.1.2.2
15		Assignment of trained and qualified personnel	4.1.2.2
16		Adequacy of staffing levels	4.1.2.2
17		Adequacy of material, equipment, space, facilities and other design and development resources	4.4.2
18	Measurement	Audit of design operations	4.17
19		Design review	4.4.6
20		Determination of measurements for product acceptance	4.11.2
21		Design verification (tests, demonstrations, analysis)	4.4.7
22		Design validation (user needs)	4.4.8
23		Selecting measuring devices	4.11.2a
24		Measurement uncertainty	4.11.1
25		Environmental conditions for measurement	4.11.2g
26		Investigation of design nonconformities	4.14.2b
27		Corrective action on designs	4.14.2c
28		Preventive action	4.14.3
29		Determination of design control system effectiveness	4.1.3
30	Personnel	Defined responsibility and authority for design personnel	4.1.2.1
31		Participants at design reviews	4.4.6
32		Understanding quality policy	4.1.1
33		Qualifications of design personnel	4.18
34		Identifying training needs for design personnel	4.18
35		Authority to identify design problems	4.1.2.1
36		Authority to initiate preventive actions	4.1.2.1
37		Authority to recommend solutions to design problems	4.1.2.1
38		Authority to verify solutions to design problems	4.1.2.1
39		Authority to control further processing of nonconforming designs	4.1.2.1
40		Authority for disposition of nonconformities	4.13.2
41	Documents	Availability of documents	4.5.2
42		Control procedures for design	4.4.1
43		Design and development plans	4.4.2
44		Organizational interface data	4.4.3
45		Technical interface data	4.4.3
46		Design input requirement	4.4.4

Table 9-6 Design control check list (continues)

Item	Factor	Topic	Clause
47		Design output documentation	4.4.5
48		Design review plan	4.4.6
49		Design verification requirements	4.4.7
50		Design validation requirements	4.4.8
51		Control procedures for use of statistical techniques	4.20.2
52		Control procedures for comparative references	4.11.1
53		Procedures for the control of design documentation	4.5.1
54		Issue of design documents	4.5.2
55		Changes to design documents	4.5.3
56		Revision control	4.5.2
57		External documents used in design and development	4.5.1
58		Obsolete documents	4.5.2
59	Data	Sources of data used in design and development	4.5.1
60		Electronic storage of data	4.5.1
61		Control of data	4.5.2
62		Changes to data	4.5.3
63	Materials	Identity of materials used to verify design	4.8
64		Marking processes	4.8
65		Inspection status	4.12
66	Equipment	Control of computer aided design/engineering tools	4.11.1
67		Calibration of measuring devices	4.11.1
68		Proving of design techniques	4.11.1
69		Development hardware/software verification	4.11.1
70		Identification of statistical techniques	4.20.1
71		Customer supplied product	4.7
72	Records	Design review records	4.4.6
73		Design verification records	4.4.7
74		Training records for design staff	4.18
75		Audit records for design processes	4.17
76		Subcontractor records of design activities	4.16
77		Design nonconformity investigations	4.14.2b
78		Comparative device verification	4.11.1

Table 9-6 Design control check list (continued)

Item	Factor	Topic	Clause
1	Product	Type of products/services produced/delivered	
2	Process	Identification of processes	4.9
3		Suitability of equipment	4.9b
4		Suitability of working environment	4.9b
5		Reference standards, codes, quality plans	4.9c
6		Process approval	4.9e
7		Equipment approval	4.9e
8		Workmanship criteria	4.9f
9		Equipment maintenance	4.9g
10		Process capability	4.9g
11		Special processes	4.9
12		Held product	4.10.3b
13		Product release	4.10.4
14		Product handling	4.15.2
15		Product preservation	4.15.5
16		Product packaging (in-process)	4.15.4
17		Work-in-progress stores	4.15.3
18	Resources	Identifying resources needed to produce products and services	4.1.2.2
19		Assignment of trained and qualified personnel	4.1.2.2
20		Adequacy of staffing levels	4.1.2.2
21		Adequacy of material, equipment, space, facilities and other production resources	4.2.3
22	Measurement	Audit of order processing operations	4.17
23		In-process inspection	4.10.3a
24		Final inspection	4.10.4
25		Inspection and test status	4.12
26		Selection of measurement devices	4.11.2a
27		Measurement uncertainty	4.11.2.1
28		Process parameter monitoring	4.9d
29		Product characteristics monitoring	4.9d
30		Environmental conditions for measurement	4.11.2g
31		Identification of statistical techniques	4.20.1
32		Inspection status	4.12
33		Product identification	4.8
34		Nonconforming product controls	4.13.1
35		Corrective action	4.14.2
36		Preventive action	4.14.3
37		Determination of process control system effectiveness	4.1.3
38	Personnel	Defined responsibility and authority for process control personnel	4.1.2.1
39		Understanding quality policy	4.1.1
40		Qualifications of process control personnel	4.18
41		Identifying training needs for process control staff	4.18
42		Authority to identify order process control problems	4.1.2.1b
43		Authority to initiate preventive actions	4.1.2.1a
44		Authority to recommend solutions to process control problems	4.1.2.1c
45		Authority to verify solutions to process control problems	4.1.2.1d

Table 9-7 Production/Installation/Servicing check list (continues)

Item	Factor	Topic	Clause
46		Authority to control further processing of nonconforming product	4.1.2.1e
47		Authority for the disposition of nonconforming product	4.13.2
48	Documents	Availability of documents	4.5.2
49		Procedures for controlling processes	4.9a
50		Handling, storage, preservation and packaging procedures	4.15.1
51		Inspection and test procedures	4.10.1
52		Process qualification requirements	4.9
53		Control procedure for statistical techniques	4.20.2
54		Product identification procedure	4.8
55		Procedure for controlling process documentation	4.5.1
56		Issue of process documents	4.5.2
57		Changes to process documents	4.5.3
58		Revision control	4.5.2
59		Obsolete documents	4.5.2
60		External documents used in process control	4.5.1
61	Data	Sources of data used in process control	4.5.1
62		Electronic storage of data	4.5.1
63		Control of data	4.5.2
64		Changes to data	4.5.3
65	Materials	Identity of materials	4.8
66		Marking processes	4.15.4
67		Inspection status	4.12
68	Equipment	Control of measuring devices	4.11.1
69		Calibration status of measuring devices	4.11.2d
70		Control of comparative references	4.11.2
71		Handling & preservation of measuring devices	4.11.2h
72	Records	Qualified processes	4.9
73		Comparative device verification	4.11.1
74		Inspection and test	4.10.5
75		Product nonconformities	4.13.2
76		Investigation of product nonconformities	4.14.2b
77		Training records for process staff	4.18
78		Audit reports for process control audits	4.17

Table 9-7 Production/Installation/Servicing check list (continued)

Procurement

Application guidance

Procurement is the name given to that area of the business that acquires materials, components, services etc. from external suppliers for use in connection with the design, development, production, installation and servicing of products and services. Procurement commences with an identified need and ends when that need has been satisfied and the supplies are made available for use by the business. Whilst some supplies will not be available for use until deposited in a store or stockroom, stores/stockrooms are covered by a separate check list as there may be different types of stores in use (see under *Support Services*). The auditor needs to establish the type of supplies that are purchased and then decide which requirement of the standard will apply. Procurement of contract labour, design services, bureau services etc. need to be treated differently from the procurement of tangible items. The check list may need to be cycled several time for different types of purchase. The procurement of significant items may require the evaluation of potential suppliers through a competitive tendering process. Separate check lists could be prepared for this activity alone.

The purpose of this check list is to establish that items from external sources will meet specified requirements prior to acceptance into the organization.

Item	Factor	Topic	Clause
1	Product	Types of procured supplies	
2	Process	Identifying procurement needs	4.2.3b
3		Purchasing specifications	4.6.3b
4		Evaluation of subcontractors	4.6.2a
5		Control of subcontractors	4.6.2b
6		Purchasing data content	4.6.3
7		In-plant surveillance by purchaser	4.6.4.1
8		In-plant surveillance by customer	4.6.4.2
9		Premature release of product	4.10.2.3
10		Handling of incoming product	4.15.2
11		Customer supplied product	4.7
12		Preservation of incoming product	4.15.5
13	Resources	Identifying resources needed to procure products and services	4.1.2.2
14		Assignment of trained and qualified personnel	4.1.2.2
15		Adequacy of staffing levels	4.1.2.2
16		Adequacy of material, equipment, space, facilities and other procurement resources	4.2.3
17	Measurement	Audit of procurement operations	4.17
18		Receiving inspection	4.10.2.1
19		Evidence of conformance provided	4.10.2.2
20		Determination of product measurements	4.11.2a
21		Selection of measurement devices	4.11.2a

Table 9-8 Procurement check list (continues)

Item	Factor	Topic	Clause
22		Measurement uncertainty	4.11.1
23		Customer supplied product problems	4.7
24		Nonconforming product controls	4.13.1
25		Inspection status	4.12
26		Product identification	4.8
27		Environmental conditions for measurement	4.11.2g
28		Identity of nonconforming product	4.13.1
29		Segregation of nonconforming product	4.13.1
30		Corrective action	4.14.2
31		Preventive action	4.14.3
32		Determination of procurement system effectiveness	4.1.3
33	Personnel	Defined responsibility and authority for procurement personnel	4.1.2.1
34		Understanding quality policy	4.1.1
35		Qualifications of procurement personnel	4.18
36		Identifying training needs for procurement staff	4.18
37		Authority to identify procurement problems	4.1.2.1b
38		Authority to initiate preventive actions	4.1.2.1a
39		Authority to recommend solutions to procurement problems	4.1.2.1c
40		Authority to verify solutions to procurement problems	4.1.2.1d
41		Authority to control further processing of nonconforming product received from suppliers	4.1.2.1e
42		Authority for the disposition of nonconformities	4.13.2
43	Documents	Availability of documents	4.5.2
44		Purchasing data	4.6.3
45		Procurement procedures	4.6.1
46		Procedure for controlling procurement documents	4.5.1
47		Issue of purchasing documents	4.5.2
			4.6.3
48		Changes to procurement documents	4.5.3
49		Revision control	4.5.2
50		Obsolete documents	4.5.2
51		External documents used in procurement	4.5.1
52	Data	Sources of data used in procurement	4.5.1
53		Electronic storage of data	4.5.1
54		Control of data	4.5.2
55		Changes to data	4.5.3
56	Materials	Identity of materials	4.8
57		Marking processes	4.15.4
58		Inspection status	4.12
59	Equipment	Control of measuring devices	4.11.1
60		Calibration status of measuring devices	4.11.2d
61		Control of comparative references	4.11.2
62		Handling & preservation of measuring devices	4.11.2h
63	Records	Acceptable subcontractors	4.6.2c
64		Inspection and test	4.10.5
65		Comparative references verification	4.11.1
66		Audit reports of purchasing audits	4.17
67		Product nonconformities	4.13.2
68		Training records for procurement staff	4.18
69		Lost or damaged customer supplied product	4.7

Table 9-8 Procurement check list (continued)

Support services

Application guidance

Support services is the name given to a range of activities within the business which support the product/servicing generation processes. They include plant maintenance, calibration, stores, library, reprographics, information technology, test laboratories, technical services, toolroom etc. They are in general off-line, in that the product or service does not pass through them for added operations before re-entering the generation processes. Where the services do interact with the deliverable product, it is usually through samples rather than the complete batch. The varieties are so great that it is difficult to create a generic check list. A stores check list is provided in Chapter 3 but even this may need to be adapted to suit the type of store. There may be a receiving store, a work-in-progress store, a finished goods store and several special stores for particular items. Since the most common support service is a calibration laboratory this has been chosen as the model; however, even this will vary depending on whether it is a mechanical or electrical calibration laboratory. Other support services check lists can be generated for specific cases.

The purpose of this check list is to establish that the provisions made for the calibration and maintenance of measuring equipment will result in devices of known accuracy and precision being available for use at all times.

Item	Factor	Topic	Clause
1	Product	Types of equipment being calibrated	
2	Process	Identification of equipment affecting product quality	4.11.2.b
3		Recall system	4.11.2c
4		Calibration frequency	4.11.2c
5		Calibration standards	4.11.2b
6		Uncommon standards	4.11.2b
7		Traceability to national standards	4.11.2b
8		Pre-adjustment checks	4.11.2f
9		Out of calibration on receipt	4.11.2f
10		Calibration process	4.11.2c
11		Post adjustment checks	4.11.2c
12		Acceptance criteria	4.11.2c
13		Unsatisfactory results	4.11.2c
14		Adjustment security measures	4.11.2i
15		Maintenance of devices	4.11.1
16		Handling, storage, packaging & preservation	4.11.2h

Table 9-9 Calibration laboratory check list (continues)

Item	Factor	Topic	Clause
17	Resources	Identifying resources needed to calibrate measuring devices	4.1.2.2
18		Assignment of trained and qualified personnel	4.1.2.2
19		Adequacy of staffing levels	4.1.2.2
20		Adequacy of material, equipment, space, facilities and other calibration resources	4.2.3
21	Measurement	Audit of calibration operations	4.17
22		Selection of measurement devices	4.11.2a
23		Measurement uncertainty	4.11.1
24		Environmental conditions for measurement	4.11.2g
25		Corrective action	4.14.2
26		Preventive action	4.14.3
27		Determination of calibration system effectiveness	4.1.3
28	Personnel	Defined responsibility and authority for calibration personnel	4.1.2.1
29		Understanding quality policy	4.1.1
30		Qualifications of calibration personnel	4.18
31		Identifying training needs	4.18
32		Authority to identify calibration problems	4.1.2.1
33		Authority to initiate preventive actions	4.1.2.1
34		Authority to recommend solutions to calibration problems	4.1.2.1
35		Authority to verify solutions to calibration problems	4.1.2.1
36	Documents	Availability of documents	4.5.2
37		Calibration procedures	4.11.1
38		Equipment maintenance procedures	4.11.1
39		Uncommon calibration standards	4.11.2b
40		Test result validity assessment	4.11.2.f
41		Procedure for controlling calibration documentation	4.5.1
42		Issue of calibration documents	4.5.2
43		Changes to documents	4.5.3
44		Revision control	4.5.2
45		Obsolete documents	4.5.2
46		External documents used in calibration processes	4.5.1
47	Data	Sources of data used in calibration services	4.5.1
48		Measuring equipment technical data	4.11.1
49		Electronic storage of data	4.5.1
50		Control of data	4.5.2
51		Changes to data	4.5.3
52	Materials	Identity of materials	4.8
53		Marking processes	4.15.4
54		Inspection status	4.12
55	Equipment	Control of primary/secondary standards	4.11.1
56		Calibration status of primary/secondary standards	4.11.2d
57		Calibration of primary/secondary standards	4.11.2b
58		Out of calibration results	4.11.2f
59		Control of comparative references	4.11.2
60		Handling & preservation of primary/secondary standards	4.11.2h
61	Records	Calibration	4.11.2e
62		Audit reports of calibration audits	4.17
63		Training records for calibration laboratory staff	4.18
64		Comparative references verification	4.11.1

Table 9-9 Calibration laboratory check list (continued)

Order fulfilment

Application guidance

Order fulfilment is the name given to a group of processes that completes the consignment/assignment or makes the delivery once the product or service has been executed. It closes the loop by delivering what the customer ordered. Fulfilment of the order may take place after dispatch of product to the customer if installation or other operations are required to be carried out. Order fulfilment ends when the customer accepts ownership of the product or when the business has been accomplished to the customer's satisfaction.

The purpose of this check list is to establish that adequate provision has been made to ensure that items ordered will be received by the customer in accordance with contract requirements.

Item	Factor	Topic	Clause
1	Product	Types of consignment (products, spares, services, assistance etc.)	
2		Inspection status	4.12
3		Product identification	4.8
4	Process	Consignment instructions	
5		Preservation	4.15.5
6		Packaging	4.15.4
7		Marking processes	4.15.4
8		Handling	4.15.2
9		Delivery	4.15.6
10		Transportation	4.15.6
11	Resources	Identifying resources needed for order fulfilment	4.1.2.2
12		Assignment of trained and qualified personnel	4.1.2.2
13		Adequacy of staffing levels	4.1.2.2
14		Adequacy of material, equipment, space, facilities and other resources	4.2.3
15	Measurement	Audit of order fulfilment operations	4.17
16		Final inspection	4.10.4
17		Selection of measurement devices	4.11.2a
18		Measurement uncertainty	4.11.1
19		Environmental conditions for measurement	4.11.2g
20		Nonconforming product controls	4.13.1
21		Corrective action	4.14.2
22		Preventive action	4.14.3
23		Determination of order fulfilment system effectiveness	4.1.3
24	Personnel	Defined responsibility and authority for order fulfilment personnel	4.1.2.1
25		Understanding quality policy	4.1.1

Table 9-10 Order fulfilment check list (continues)

Item	Factor	Topic	Clause
26		Qualifications of order fulfilment personnel	4.18
27		Identifying training needs	4.18
28		Authority to identify order fulfilment problems	4.1.2.1
29		Authority to initiate preventive actions	4.1.2.1
30		Authority to recommend solutions to order fulfilment problems	4.1.2.1
31		Authority to verify solutions to order fulfilment problems	4.1.2.1
32		Authority to control further delivery of nonconforming product	4.1.2.1
33		Authority for the disposition of nonconformities	4.13.2
34	Documents	Availability of documents	4.5.2
35		Handling procedures	4.10.1
36		Packaging procedures	4.10.1
37		Preservation procedures	4.10.1
38		Delivery procedures	4.10.1
39		Consignment documentation	4.10.5
40		Procedure for controlling order fulfilment documents	4.5.1
41		Issue of process documents	4.5.2
42		Changes to documents	4.5.3
43		Revision control	4.5.2
44		Obsolete documents	4.5.2
45		External documents used in order fulfilment	4.5.1
46	Data	Sources of data used in order fulfilment	4.5.1
47		Electronic storage of data	4.5.1
48		Control of data	4.5.2
49		Changes to data	4.5.3
50	Materials	Identity of marking materials	4.8
51		Marking processes	4.15.4
52		Inspection status of marking materials	4.12
53	Equipment	Control of measuring devices	4.11.1
54		Calibration status of measuring devices	4.11.2d
55		Control of comparative references	4.11.2
56		Handling & preservation of measuring devices	4.11.2h
57	Records	Inspection and test	4.10.5
58		Audit reports of order fulfilment audits	4.17
59		Product nonconformities	4.13.2
60		Training records for order fulfilment staff	4.18

Table 9-10 Order fulfilment check list (continued)

Chapter 10

Tips for auditors

This chapter provides hints, tips and guidance for auditors not included elsewhere in the book. Whilst the auditing process can be fairly uniform from one organization to another, each audit is a new experience. The organization will differ in the manner in which it organizes resources and the method of executing work. Interpretation of the standard will differ depending on the circumstances and even when dealing with organizations that perform the same processes as others, something will be different apart from the people. There is no right or wrong way of doing something just more or less effective ways. Just because an organization uses a novel method does not mean it is ineffective or noncompliant. The guidance is presented in alphabetical order for ease of reference.

Acceptance criteria

❏ At stages in the process where decisions are made check the criteria being used, as often, informal standards, limits and acceptance criteria are used to accept or reject work or decide a course of action. The people producing the work should be aware of the acceptance criteria, otherwise there is no legitimate basis for quality control and control becomes the domain of the powerful.

Auditing approach

❏ External audits should be performed against the external standards/contracts, not the internal procedures.

❑ Audit areas, not clauses of the standard; i.e. examine processes against the clauses of the standard which apply to the processes. Do not audit clause by clause. Companies are not organized like that.

- Know what you want to establish.

- Gather information relative to what you want to establish.

- Seek conformity not nonconformity.

- Don't follow a trail unless it is pertinent to your objective.

- Record your findings.

- Check that you have obtained the objective evidence needed to prove you have achieved your objective before moving on.

- Praise good points and don't criticize bad points.

- Thank all concerned.

❑ Be prepared to accept that you may be wrong or mistaken.

Availability of personnel

❑ Check who will attend the Opening Meetings and Closing Meetings and whether the managers you expect to interview are available *before* you arrive on site.

Calibration

❑ Not all measuring equipment requires calibration. First establish what it is being used for, then apply the relevant requirements.

❑ Check that the specific devices used for measurement are recorded, as otherwise the company will not be able to assess the effects of invalid verification results when the devices are subsequently found to be out of calibration.

❑ Check what companies do if calibration standards as well as the working standards are found to be inaccurate by internal/external test laboratories.

❑ Do remember that periodic calibration is only necessary when it is possible for the accuracy and precision of the measuring device to vary. Steel measuring tapes do not need periodic calibration. They do need periodic inspection for damage.

❑ Check how the calibration frequency has been determined and if it remains fixed regardless of use and environment, the system may be ineffective.

Commitment

❑ Check that agreements to corrective action are honoured both from internal audit results and other sources of corrective action.

❑ Check records of concessions on outgoing product to establish whether the commitments of upper management are being honoured.

❑ Establish the methodology for developing the quality system. If it is based on the 'document what you do and do what you document' principle then verify that the necessary controls have been put in place and people are implementing the new practices. If it is based on the 'top down design' principle, verify that the ideas presented in the documentation are actually being followed. In both cases verify that the staff concerned were involved in the development of the system.

❑ If staff are committed to quality they will not tolerate errors and recurring problems, they will attempt to remove the flaws in the system.

❑ If managers are committed to quality they will act upon being notified of errors and recurring problems and authorize changes to the system.

Concessions

❑ Do check that the time limitation of concessions or waivers has not been exceeded.

❑ Check that concessions are not used to process document changes as this is an illegitimate use of the concession.

Conduct

❑ Don't load your questions with the expected answer. Ask open questions that allow the person to talk freely.

❑ Do not get angry or raise your voice to get a message across. Change the direction of your questions if they are not achieving the desired result.

❑ Don't cause a change in the performance of staff by your actions.

Confirming the facts

❏ Don't move onto new areas of investigation until facts have been confirmed and objective evidence has been acquired to support them.

❏ Don't announce you have confirmed that a nonconformity exists before you have assessed its significance.

Contract amendments

❏ Verify that amendments to contract are approved by the same functions or organizations as approved the original contract and that amendments are conveyed to those who will implement them.

Contract review

❏ Don't assume that all customers place written orders, many don't but the company has to document them. Check that those accepting orders know of the companies capability to meet the specific requirements.

Contract review records

❏ These may be minutes of a meeting, a completed check list containing relevant questions or simply a signature on an order indicating it has been accepted. If the latter, verify that those making the acceptance decision have access to data defining the company's capabilities, such as a database containing stock and delivery data.

Contracts

❏ Ask to examine a sample of typical contracts so that you determine if the quality system is capable of enabling the company to meet such requirements.

❏ Do check the distribution of contracts/orders and verify that those purchasing product have access to contract data in order to pass relevant data onto suppliers. Check that those releasing product to customers have access to contract data in order to verify that requirements have been fulfilled prior to shipment.

❏ Verify that the company has a means of conveying non-standard conditions to the point where they will be implemented.

Control charts

❏ Establish whether staff know what the dots on the charts mean. If they don't then they cannot claim to be in control of the process.

Controls

❏ Controls regulate outputs to ensure standards are maintained. Auditors should therefore establish that the controls installed by the company are effective in maintaining conformity with the prescribed standards. To do this, the auditor needs to establish that processes provide staff with:

- Knowledge of what they are required to achieve
- Knowledge of what they are achieving
- The ability to judge conformance with the defined standards
- Authority to change their performance should standards not be met

❏ Do check downstream of the operation being audited to gather facts on the effectiveness of the controls.

Corrective and preventive action

❏ Corrective action is action taken to prevent the recurrence of a nonconformity, i.e. a nonconformity has to exist for corrective action to be possible.

❏ Preventive action is action taken to prevent the occurrence of a nonconformity.

❏ Action taken to correct a nonconformity or eliminate its cause is a remedial action.

❏ Do check that actions are placed on staff who have the authority to take the specified action, otherwise timely action will not be taken.

❑ Do check that provisions have been made to monitor the implementation of actions and that they do not rely on feedback from the implementor, who may place low priority on the action.

Customer complaints

❑ When checking the file of customer complaints look at what remedial action has been taken but also check if any corrective action has been taken (i.e. action to prevent a recurrence).

Customer supplied product

❑ Always ask if there is any customer supplied product before examining the adequacy of the controls.

❑ Establish where the customer supplied product is stored since if no one knows where it is, it cannot be claimed to be under control.

Decisions

❑ Do establish the criteria that staff employ to make decisions since if this varies for the same type of decision, the quality of output will vary accordingly.

Departmental procedures

❑ Beware of departmental procedures as they often neglect to define the inter-faces between the departments that supply input products/services or receive output products/services.

Design

❑ Whilst not covered by the standard, the design activity between design input and design output needs to be controlled. Some auditors apply the requirements of clause 4.9, but it is more logical to apply basic control theory. For example, whatever is done:

- What is the requirement?
- How is it conveyed to the design staff?

- What are the acceptance criteria?

- What standards and guidelines ensure consistency, interchangeability etc.?

- How were these standards approved or validated?

- What checks are performed to verify that standards are maintained?

- How are results conveyed to design staff?

- How are changes implemented if errors are found?

- What records are kept to demonstrate that standards have been maintained?

❑ Where the product requirements are defined in performance terms then design is required to convert customer needs into a definitive specification. So investigate the nature of the company's design requirements and where they come from.

Design authority

❑ Do check that those changing the design have the authority to do so and that any specialist groups have been consulted.

Design changes

❑ Control of the design should commence immediately once the design input requirements have been agreed. Thereafter, all changes should come under configuration control.

❑ Verify that previous analytical studies and test results are evaluated whenever there is a design change.

Design input

❑ Remember that a design input document is one which relates to the requirements of the standard and hence is governed by element 4.5. It has to be approved before issue.

Design reviews

❏ Remember that design reviews are reviews of a design not a progress meeting. The review should precede a formal meeting. As with management reviews, the design review meeting should examine the result of the review in order to draw conclusions and decide on corrective actions.

❏ Check that the attendees are given design data before the meeting.

Design status

❏ Do check that the status of the design is recorded in any analytical studies, otherwise their validity to later designs cannot be confirmed.

Design verification

❏ Verify that there is traceability between all design requirements and the records of design verification. Often such records only contain the results of verification tests and do not include mathematical analysis, alternative calculations, predictions etc.

Document approval

❏ Don't assume that every document must have a signature. The method of approval may require a separate approval sheet.

Document maintenance

❏ Remember that to maintain a document means to keep it current so you need to establish what provisions the company has made to keep its documentation up to date.

Document revision

❏ Don't assume that revision status must always be a number or letter. It can be by date.

❏ Watch out for pages extracted from procedures and posted close to workstations. Unless the extra copies are issued with the procedure from which they have been extracted they may be obsolete.

Documentation audit

❏ Don't request a copy of the documented quality system as you may be surprised at what you receive. Ask firstly 'What constitutes your documented quality system?' and then request particular documents, such as a copy of the policy manual, a procedures list and a sample of procedures.

Documentation control

❏ Verify that all the documents that are used to achieve quality are under document control, not just the quality system procedures, as is often the case.

Document changes

❏ Verify that in approving a change to a document, the effects of the change on other documents are considered, otherwise the system may not be being maintained.

❏ Establish what mechanism is used for changing a range of documents, as often the costs of change are excessive and informal methods may be used for the one-off circumstances.

Electronically-stored documents

❏ Check read/write authorization, password, storage, computer virus and copying security.

❏ Check which version is the master: the hard copy or the electronically-stored copy.

Forms

❏ Don't pursue unfilled boxes on forms that do not serve the achievement of product/service quality.

❏ All forms should be traceable to the control procedures that govern their applicability.

❏ All forms are not records and all records are not quality records.

Hearsay evidence

❏ Do not accept hearsay evidence. Take a note and, if time permits, check the facts.

Informalities

❏ Constantly look for informal activities and evidence that without them the process, product or service would be nonconforming. Informalities that don't affect quality of product, process or service are merely interesting.

❏ Look for staff doing things for which there is no requirement. Some of these things may affect quality. The others may be depleting resources and causing errors to escape detection.

Internal audit

❏ Verify that the internal audit programme addresses the system as a whole, not just the procedures. Internal audits should address the organization structure, the processes and the resources for the audit to be claimed as an audit of the quality system.

❏ Establish the means by which the organization determines the effectiveness of the system (see under *Quality system effectiveness*).

❏ Establish that the audit programme is adjusted to take account of new practices, new processes, new technology and changes in the organization structure.

Job descriptions

❏ The standard does not require job descriptions. Ask: 'In what document are your responsibilities and authority defined?'

❏ Check that such documents, if the policy requires them, define responsibilities, authority and accountabilities.

Laws and regulations

❏ Establish what local and industry sector laws and regulations apply to products, services and the operations of the company before commencing the audit. Customers do not expect organizations to be closed down by failure to abide by the law.

Maintaining policy

❏ Maintaining a policy means more than keeping a document up to date. Check that there is a means for regulating deviations from the policy and for ensuring that it remains compatible with changing customer needs and expectations.

Maintaining the system

❏ System maintenance is more than keeping documents aligned with practices. The documents should be changed before the practices, otherwise the system as documented is not being implemented. Verify that changes in the organization, in markets, in technology etc. cause a review of the system to establish their effects.

❏ System maintenance also means that the effects of document changes on other documents are taken into account when approving a change to one of the system documents.

Management Representative

❏ The Management Representative can be anyone with the appropriate authority. The post need not be a full-time job and can be shared by a person having other positions, which may appear to create conflict. Unless evidence is found where the person is clearly biased towards the goodness of the operations they control, then one has to give the company the benefit of the doubt.

❏ In checking these requirements, the auditor should verify that the Management Representative is empowered to make decisions on the design, implementation, evaluation and improvement of the quality system. The auditor should check that this person has control over the quality system documentation and any changes.

The Tale of the Hastily Prepared Report

The team had not had enough time to prepare their report but Charlie felt they had done a good job and called in George, the Management Representative, to present their findings. Charlie went through the nonconformities. There were 5 majors and 38 minors. George was somewhat shaken by the result and could not believe there were so many major problems as he had been confident that the system met the requirements of ISO 9002.

Charlie presented the major nonconformities first. There were 3 instances of equipment being overdue for calibration, 2 instances of training records not being available, 1 instance of a delivery schedule not being approved, 1 instance of an unapproved subcontractor being used and 1 instance of there being no review of tooling design.

George felt he understood ISO 9002 and remembered the definitions that Charlie had stated at the Opening Meeting so was prepared to challenge the majors.

George questioned the one on calibration, reminding Charlie that he had said that a major was a total breakdown of the system. He remarked that since only three instruments had been found overdue, it hardly demonstrated that the system had broken down when they had over 200 instruments in the calibration system. Charlie agreed to change it to a minor nonconformity.

On the question of the unavailability of training records, George asked how that could be a major when the requirement for training records is only one requirement of a clause in the standard. George distinctly remembered Charlie saying that a major was a failure to meet the requirements of a clause in the standard and understood from Charlie that it meant all requirements in a clause. After some careful thought Charlie agreed.

George's confidence was now increasing and he felt he could challenge the others successfully. The nonconformity concerning tooling design was a mystery to George and he asked Charlie where tooling design was addressed in ISO 9002. Charlie claimed that it was covered by element 4.4 but when George pointed out that these requirements apply to the products they sell and they don't sell tooling, he had to agree that the requirements were not relevant in this case.

George now having sent Charlie in full flight, tackled him on the unapproved subcontractor and the unapproved delivery schedule. He succeeded in reducing two to a minor and eliminating the third entirely, as delivery schedules are not documents related to ISO 9002.

The moral of this story is:

More haste less credibility!

Management responsibility

❏ There are no procedures required for the management responsibility element of ISO 9001.

❏ Check that statements of responsibility also include the discretion that staff have in making decisions. Job descriptions often omit this aspect.

Management review

❏ The records of the review should contain the results of the review, as well as the results of the meeting held to discuss the results of the review.

Measuring equipment

❏ Always ask what an instrument is used for before examining the adequacy of controls.

Modifications

❏ Modifications are changes to hardware made as a result of changes to a design so check the company has a controlled mechanism for changing all hardware affected by the design change, including product in production, in stores and in service.

❏ Establish how modifications are proven. If production models are used, then verify that adequate controls exist to prevent nonconforming product re-entering the production stream.

New practices

❏ Establish how new practices are introduced. Does the document change follow the change in practice or vice versa? Changes in practice ought to follow a change to the documentation otherwise the documented system is not being implemented.

❏ Verify that staff receive instruction when changes to the documentation affect what they do.

❏ Establish whether the audit programme is adjusted to take account of new practices.

Notices

❏ Check than any notices posted in the area are relevant, current and controlled if affecting product quality. An obsolete 'Zero Defects' notice can send out the wrong message when judged against current policies. Check that the notices do not conflict with the practices documented in the procedures.

Objective evidence

❏ The standard does not require that compliance with every requirement be demonstrated on the day of the audit. Beware of those lean systems that do not store endless quantities of information just waiting for the auditor to ask a pertinent question. You may have to return at a more appropriate time to observe an operation being carried out and so obtain the necessary objective evidence.

Objectives

❏ Objectives remain in force until achieved whereas policies remain in force until changed. Quality objectives are not the same as quality policies.

❏ Be clear about your objectives before commencing an interview or entering an area to commence auditing. Ask questions which will produce evidence that is relevant to the achievement of your objectives.

Obsolete documents

❏ Establish whether staff are aware that certain documents are obsolete and what they would be used for before examining the controls. Controls are needed to prevent the undesirable. Establish what undesirable event could happen before reaching your conclusion.

Organization charts

❏ Organization charts can be used to depict interrelationships of personnel but they often do not show relationships across department boundaries. They are therefore not complete evidence of compliance with the requirement for interrelationships to be defined and documented.

❏ Organizations charts are a solution to a requirement or need but in themselves are not normally a statement of policy. However, one has to establish the purpose that is being served by their inclusion in a quality manual, as they may in fact be stating policy.

❏ Organization charts often show staff relationships but only the chain of command. Establish how cross-departmental communication takes place and whether it is effective.

❏ Organization charts should be controlled documents if they document staff interrelationships, but establish whether this is so before judging that the display of obsolete charts is a nonconformity. Place it as a low priority in any case as it does not affect product quality.

Organizational and technical interfaces

❏ Many auditors regard the requirement addressing organizational and technical interfaces as applying only to the organization when in fact it also applies to technical data. Auditors should verify that the company has made provision for controlling technical interfaces between products and services.

Planning the audit

❏ Don't rely on one contact name as anything could happen to cause he/she to be unavailable prior to the audit, even on the day.

❏ Do establish the dress code since to turn up in smart business attire may be intimidating to staff who always wear casual clothes. Likewise, the pony tail, jazzy tie, pink suit etc. can be off putting. Try to avoid distractions.

❏ Do establish the names of the key managers that need to be available and communicate this to the organization, otherwise the may have taken the day off.

Plans

❏ ISO 9001 requires very few formal plans to be produced. It requires consideration to be given to preparing quality plans, requires design and development plans and for production, installation and services processes to be planned. Where the auditor detects problems and discovers that operations have not been planned, then if specific planning requirements of the standard are deemed not to be applicable, clause 4.14.3b on preventive action can always be applied.

Preventive action

❏ Whilst the standard is aimed at preventing nonconformity, the auditor has to consider whether it is reasonable to expect in the circumstances that preventive action is necessary. For example, records may be lost if the door to the room in which they are stored has no lock, but if on carrying out a sample check, no records are missing, then it can be assumed that the supplier's storage provisions for records are adequate. The fact that someone may enter the room and steal some records is irrelevant as they could do this even if there was a lock. Ask if they have had any difficulty locating records. The answer may indicate whether the provisions they have made are adequate.

❏ Clause 4.14.3b is possibly the most important clause in ISO 9001 and can be applied to any situation where the organization has not planned to prevent problems.

Procedures

❏ Don't expect a procedure for every task. Procedures are documents that define a sequence of tasks to execute a routine activity so not all quality system documents will be procedures, some will be standards, guidance documents, forms etc.

❏ Remember that activities which require skill and judgement cannot be proceduralized and that the requirement in the standard for procedures is plural, not singular in all cases.

❏ Do establish what constitutes a 'documented procedure' since assumptions here may again lead to conflict.

❏ Do verify that procedures are traceable to policies and vice versa.

Production

❏ Verify that the design being reproduced is one that has been proven and that evidence exists to support such a conclusion.

❏ Verify that production equipment and tooling has been proven before production commences as it is a preventive action covered by clause 4.14.3.

Prototypes

❏ Prototype models used to prove a design should be representative of production standards and need to be controlled such that the standards to which they are built can be demonstrated.

Publicity literature

❏ Often overlooked as having no affect on quality, publicity literature sets customer expectations and if not reflected in the product or service leads to unsatisfied customers.

Purchasing

❏ Establish what it is that the company purchases before examining how they purchase items and don't forget about the intangibles such as the purchase of subcontract labour, consultancy services etc.

❏ Before attacking their approved supplier list, establish what their procurement policy is and then judge accordingly.

Qualified personnel

❏ Qualified personnel are personnel whose qualifications match those required for a particular job. Qualifications include education, training, experience and personal attributes.

Quality Manuals

❏ Do explore generalities in the Quality Manual as they often are no more than words with no method of implementation in place.

❏ Don't judge there to be nonconformities solely because no evidence is available. Examine the internal audit reports to establish that the operations you are interested in have been audited.

❏ If the provisions for quality records appear somewhat brief, determine whether they meet the criteria of the ISO 8402 definition.

Quality planning

❏ Every product or service is different so the auditor should verify that the company has made provision for defining specific product/service requirements that differ from the norm and for conveying these requirements to those who will implement them. This is the function of the quality plan.

Quality policy

❏ The quality policy statement should be a controlled document but if obsolete copies are displayed then establish whether staff know the current policy before judging it to be a nonconformity.

❏ In checking the quality policy requirements, the auditor should verify that the company has a means for determining customer expectations and organizational goals even though there is no specific requirement since the adequacy of the quality policy cannot be verified otherwise. These should be among the first questions to be asked.

Quality system development

❏ There are no requirements in the standard governing the method of quality system development. However, there are requirements for quality planning in clause 4.2.1 and so there should be a list of documents to be prepared, resources to be acquired and activities that need to be carried out to complete the development of the system. There is also a requirement in clause 4.14.3b for preventive action.

Quality systems

❏ Quality systems should be designed, not constructed out of existing practices as is often the case. Effective systems are designed to fulfil a specific series of functions and new practices designed or selected to deliver the right performance.

❏ Constructing a quality system by documenting what you do merely formalizes current practice. It does not create anything new. If there is no existing system but a random collection of activities that by chance cause products to meet customer requirements, then documenting such activities merely formalizes them. It does not create a system. It only provides a system that delivers current performance.

❏ A system is an ordered set of ideas, principles and theories or a chain of operations that produces specific results and which work together in regular relationship. Therefore a quality system should be the integration of interconnected business processes that collectively cause the supply of conforming product/service and prevent the supply of nonconforming product/service.

Records

❏ Undated records are of little value as objective evidence for events.

❏ All records should have a defined location where they can be accessed for analysis.

❏ Records should be initiated only as a result of an instruction in a procedure otherwise they represent informal practices.

❏ The retention times for records can be as short as the company deems necessary for them to fulfil their purpose.

❏ Do establish what constitutes a 'quality record' and match that with the definition given in ISO 8402, in which it states that a record is a document which furnishes objective evidence of activities performed or results achieved and a quality record provides objective evidence of the extent of the fulfilment of the requirements for quality.

❏ Don't treat all records as quality records. For instance, a list of documents is not a quality record as defined in ISO 8402.

❏ Don't accept minutes of a progress meeting as a quality record unless they also meet the criteria in ISO 8402.

❏ Do check that the procedures cause the right data to be recorded, otherwise each record will vary in its content and create gaps in data when attempts are made to analyse trends.

❏ Records need to indicate who made the decisions so that the decisions can be audited. It is of less importance to know who made a decision than what criteria they used to make the decision. Errors usually result from inadequate criteria to ensure consistency in decision making.

Reporting

❏ Do keep a focus on effectiveness rather than on rigid compliance with the standard.

❏ Don't use hearsay evidence as valid evidence of nonconformities. The person giving the information may be mistaken or the information out of date.

❏ Don't report nonconformities that cannot be traced to a requirement of the standard or contract against which the audit is being conducted.

❏ ISO 9004 cannot be used to justify a nonconformity. However, divergence from ISO 9004 guidance can be used to substantiate an observation which the supplier can choose to ignore if they can put forward equally valid alternatives. Ignoring ISO 9004 guidance without alternative solutions would indicate that the supplier is not serious about managing the achievement of quality effectively.

Resources

❏ Verifying the adequacy of resources can only be accomplished through observation of resource utilization rather than through particular questions. A backlog, shortage of space, personnel or equipment should become evident when one enters an area.

Responsibilities and authority

❏ Do probe the authority of personnel as well as their responsibilities, as responsibility is often assigned without the corresponding authority being delegated.

❏ When observing signatures on documents without a printed name and position below, or other means for tracing the legitimacy of the signatory, ask how staff establish that the signatory is authorized and seek evidence of that person's documented authority to verify compliance with clause 4.1.2.1 of ISO 9001.

❏ In checking these requirements, auditors should verify that staff have the authority to do what they are found be doing and that they are doing the things for which they have responsibility.

Signatures

❏ Signatures on documents need to be traceable to the person they represent, otherwise it cannot be verified that the person signing the document had authority to do so under clause 4.5.2 and clause 4.5.3 of ISO 9001.

Starting an interview

❏ Don't start by asking people what they do. You ought to know who you want to speak to and what you wish to establish and so your questions should be more specific. In some cases 'What do you do?' questions may be appropriate but beware that you may waste time in waiting for the person to get around to what you are particularly interested in.

Supplier selection

❏ In the final analysis most companies will select suppliers on price. Check that where this is the case, the quality of competing products is the same, i.e. all meet the technical requirements and that the company has equal confidence in all to meet their requirements. Where the confidence varies, check that the company employs a method to compensate for the lack of confidence.

System effectiveness

❏ Before reaching conclusions as to the effectiveness of the quality system, look for evidence that:

- The quality system fulfils the purpose for which it was designed.

- Reported customer complaints are declining or have declined and remained at a level where variation is stable.

- Reports of internal and external nonconformities are declining or have declined and remain at a level where variation is stable.

- Process outputs connect with process inputs and that such processes are adequately defined by the documented policies and procedures.

- The organization's documented policies and procedures cause the witnessed activities to be performed.

- The organization's documented policies and procedures cause the documentary evidence to be created as seen.

- The documented policies and procedures are updated before or promptly after changes in practice.

- Trends of substandard performance are promptly corrected.

- Staff understand the documented policies and procedures.

- Problems known to staff are resolved through use of the documented policies and procedures.

- Adequate resources have been provided for those activities that affect quality.

- Staff are cognisant of the responsibilities and authority assigned to them.

- There are no gaps in or overlapping of responsibilities and authority.

- The work undertaken by staff is that required to satisfy the organization's documented policies and procedures.

- The audit programme detects problems and secures their correction before operations that may adversely affect quality continue.

- The manner in which work is organized does not result in undue delays.

Technology

❏ Check that the technology employed is appropriate to the task and is capable of meeting the declared objectives. If not obvious from what you observe, ask if they know its capability and have evidence to show that such capability is within the range required for the job.

Terminology

❏ ISO 8402 is a normative reference in ISO 9001 and is therefore mandatory and so can be used by auditors to extend the requirements of ISO 9001 but only where terms used in ISO 9001 are defined in ISO 8402.

❏ Do establish what the specified requirements mean to the organization concerned, as assumptions that are not validated may cause conflict.

❑ Clarify terminology at the outset or immediately there appears to be a misunderstanding, so as to demonstrate your openness and flexibility.

Trained personnel

❑ Trained personnel are not necessarily competent personnel. The training has to be relevant to the current job and there has to be some evaluation made to establish that the individual acquired the skills for which they were being trained.

Understanding quality policy

❑ Testing the understanding of the quality policy will not be achieved through asking people if they know the quality policy. One has to establish whether they understand what it means. So ask: 'How does the quality policy affect what you do?'

Urgencies and emergencies

❑ Find out what happens when orders are urgent or when an emergency situation arises, as it could reveal evidence that the system is bypassed.

Appendix A

Cross reference with ISO 10011

ISO 10011 is in three parts. It establishes basic audit principles and provides guidelines for planning, conducting and reporting quality system audits, managing quality system audit programmes and defines qualification criteria for quality system auditors. The standard specifically applies to quality system audits and does not apply to product audits, service audits and process audits. This book amplifies and extends the provisions of ISO 10011 but not in the order presented in ISO 10011. This appendix cross references clauses of ISO 10011 with chapters and headings in this book.

ISO 10011 Part/Clause	Title	Covered in Appx C	Handbook Chapter	Section Heading
Part 1				
3.1	Definition of quality audit	Yes	1	Purpose of audits
3.2	Definition of quality system	Yes	1	Quality system audits
3.3	Definition of quality auditor	Yes	1	Auditor, Assessor, Auditee, Assessee
3.5	Auditee	Yes	1	Auditor, Assessor, Auditee, Assessee
3.6	Observation	Yes	5	Observation statements
3.7	Objective evidence	Yes	4 5	Seeking objective evidence Documenting the audit findings
3.8	Nonconformity	Yes	3 5	The success criteria Writing nonconformity statements

ISO 10011 Part/Clause	Title	Covered in Appx C	Handbook Chapter	Section Heading
4.1	Audit objectives	Yes	1	Aims and objectives of audits
4.2	Roles and responsibilities		2	The role of auditors
4.2.1.1	Audit team		2	The audit team
			3	Audit team selection
4.2.1.2	Auditor's responsibilities		2	Auditor's role
				Auditor's responsibilities
				Auditor's authority
4.2.1.3	Lead auditor's responsibilities		2	Lead auditor role
				Lead auditor responsibilities
				Lead auditor's authority
4.2.1.4	Independence of the auditor		3	Audit team selection
4.2.1.5	Auditor's activities		2	Auditor's responsibilities
				Auditor's authority
4.2.2	Client		3	Company's responsibilities
4.2.3	Auditee	Yes	3	Company's responsibilities
5.1.1	Audit scope	Yes	3	Clarifying purpose and scope of the audit
5.1.2	Audit frequency		2	Audit frequency
5.1.3	Preliminary review of quality system description	Yes	1	Documentation audit
			3	Examination of quality system documents
			4	Documentation audit
5.2	Preparing the audit		3	Planning audits
5.2.1	Audit plan	Yes	1	Planning and organizing the audit
			3	The audit schedule
5.2.2	Audit team assignments		3	Assigning auditors
			3	The audit schedule
			3	Check lists
5.2.3	Working documents		3	Preparation check list
			3	Check lists
			5	Figures 5-3 to 5-7
5.3	Executing the audit		1	Conducting the audit
5.3.1	Opening meeting	Yes	1	Conducting the audit
			4	The opening meeting
5.3.2	Examination		1	Conducting the audit
			4	The implementation audit

ISO 10011 Part/Clause	Title	Covered in Appx C	Handbook Chapter	Section Heading
5.3.2.1	Collecting the evidence		1	Conducting the audit
5.3.2.2	Audit observations	Yes	1	Conducting the audit
			4	Recording results
			5	Documenting audit findings
5.3.3	Closing Meeting	Yes	1	Reporting the audit
			5	The Closing Meeting
5.4	Audit documents		1	Reporting the audit
			5	Reporting the audit
5.4.1	Audit report preparation		5	Drafting the audit report
5.4.2	Report content		5	Report content
5.4.3	Report distribution		5	Distribution of audit report
5.4.4	Record retention		5	Endorsement of Quality Manual
6	Audit completion		1	Reporting the audit
			5	The Closing Meeting
7	Corrective action follow-up	Yes	1	Following-up the audit
			5	Corrective action statements
			6	Quality system surveillance
Part 2				
4	Education		2	Education of auditors
5	Training		2	Training of auditors
6	Experience		2	Experience of auditors
7	Personal attributes		2	Personal characteristics of auditors
8	Management capabilities		2	Experience of auditors
9	Maintenance of competence		2	Experience of auditors
10	Language		2	Selection of auditors
11	Selection of lead auditor		2	Selection of auditors
A.2	Evaluation panel		2	Auditor evaluation
A.3	Evaluations		2	Auditor performance evaluation
Part 3				
4	Managing an audit programme		2	Audit management
4.1	Organization		2	Organization of auditors
4.2	Standards		2	Audit standards
4.3	Qualifications of staff		2	Selection of auditors

ISO 10011 Part/Clause	Title	Covered in Appx C	Handbook Chapter	Section Heading
4.3.1	Audit programme management		1	The audit programme
			2	The audit programme
4.3.2	Auditors		2	Selection of auditors
4.4	Suitability of team members		2	Selection of auditors
4.5.1	Performance evaluations		2	Auditor performance evaluation
4.5.2	Consistency of auditors		2	Performance standard
4.5.3	Training		2	Training of auditors
4.6	Operational factors		2	Audit procedures
4.7	Joint audits			
4.8	Audit programme improvement		2	Improvement
5	Code of ethics	Yes	2	Code of conduct

Appendix B

ISO 9000 statistics

ISO 9000 registered organizations

The figures in Table B-1 have been taken from the Mobile Survey and a full version can be obtained free of charge from Dr John Symonds, Technical Advisor, TQM, Mobil Europe Limited, Mobil Court, 3 Clements Inn, London WC2A 2EB. Tel +44 (0)171 412 4897; Fax +44 (0)171 412 4152.

26 certification bodies participated in the survey and 17 of these have awarded 69.7% of all certificates. The figures should not be considered accurate, as the survey has not been exhaustive. Not all certification bodies world-wide took part. There is as yet no international body that requires such information to be collected.

Region	Jan-93	%	Sept-93	%	June-94	%	Mar-95	%
World-wide	27,824	100	46,702	100	70,517	100	95,476	100
Europe (including UK)	23,092	83	37,779	80.89	55,400	78.56	71,917	75.32
Australia & New Zealand	1,862	6.67	3,184	6.82	4,628	6.56	6,479	6.79
North America	1,185	4.24	2,589	5.54	4,830	6.85	7,244	7.59
Americas	39	0.14	156	0.33	534	0.76	873	0.91
Far East	683	2.45	1,583	3.39	3,091	4.38	5,979	6.26
Rest of the world	963	3.5	1411	3.03	2,034	2.89	2984	3.13

Table B-1 Registrations by region

The Top 40 countries

In Table B-2, the top 40 countries with the most ISO 9000 registrations are listed. Again the figures are take from the Mobil Survey but as the number of certification bodies is increasing world-wide, it has not been possible to obtain data from all of them. For consistency the same bodies were contacted in the 1993 survey as in the 1995 survey. It is possible however, that undercounting has occurred, especially in Germany.

Position	Country	Regs	Position	Country	Regs
1	UK	27810	21	Austria	667
2	USA	5954	22	Finland	646
3	Germany	5875	23	Malaysia	628
4	Australia	5299	24	India	585
5	France	4277	25	Hong Kong	551
6	Netherlands	4198	26	Brazil	548
7	Italy	3146	27	Israel	497
8	Japan	1827	28	South Korea	390
9	South Africa	1627	29	China	285
10	Switzerland	1520	30	Turkey	270
11	Ireland	1410	31	Portugal	257
12	Canada	1290	32	Greece	162
13	Belgium	1226	33	Mexico	145
14	Denmark	1183	34	Hungary	125
15	New Zealand	1180	35	Czech Republic	101
16	Taiwan	1060	36	Thailand	95
17	Singapore	1003	37	Philippines	79
18	Spain	942	38	Venezuela	56
19	Sweden	871	39	Saudi Arabia	53
20	Norway	679	40	Colombia	51

Table B-2 Registrations by country

ISO 9001 statistics

In Table B-3 some statistics based on ISO 9001 are presented. The single 'shall' omitted from ANSI/ASQC Q9001 is in clause 4.2.2. The number of requirements is conditional on the manner in which they are counted (see Chapter 7).

	Characteristic
8	Conditional procedures
10	Mandatory procedures
20	Types of quality records
20	Elements
59	Clauses
138	'Shall' statements (137 in ANSI/ASQC Q9001-1994)
184	'Shall' statements if lists are included
323	Requirements

Table B-3 Characteristics of ISO 9001

The procedures and records referred to in Table B-3 are listed in Tables B-4 and B-5. It could be argued that even some of the mandatory ones are conditional. The distinction is perhaps academic; however, the tables do illustrate that options are available.

1 Contract review	6 Corrective and preventive action
2 Document and data control	7 Handling, storage, packaging, preservation and delivery
3 Inspection and testing	
4 Control, calibration and maintenance of inspection, measuring and test equipment	8 Quality records
	9 Internal quality audits
5 Control of nonconforming product	10 Training needs

Table B-4 Mandatory procedures

1 Design control	Conditional upon the company performing product or service design
2 Purchasing	Conditional upon the company purchasing product or service that affects output product or service
3 Control of customer supplied product	Conditional upon the company receiving product from customers
4 Product identification	Conditional upon procedures being necessary to define identification methods
5 Traceability	Conditional upon traceability being a condition of contract, order or legislation
6 Production, installation and servicing	Conditional upon situations where their absence could adversely affect quality
7 Control of servicing	Conditional upon servicing being a requirement of the contract or order
8 Statistical techniques	Conditional upon a need for statistical techniques being identified

Table B-5 Conditional procedures

QS9000-1994 Statistics

The additional elements within QS9000 are as follows:

1 Production part approval process
2 Continuous improvement
3 Manufacturing capabilities

The additional clauses are:

4.1.2	Organizational Interfaces
4.1.4	Business plan
4.1.5	Analysis and use of data
4.1.6	Customer satisfaction
4.2.3	Advanced Product Quality Planning Use of cross functional teams Feasibility Reviews Process Failure Mode and Effects Analysis Control Plans
4.4.2	Required skills
4.4.4	Design input supplemental
4.4.5	Design output supplemental

4.4.7	Design verification supplemental
4.4.9	Design changes supplemental
4.5.1	Reference documents Document identification for special characteristics
4.5.2	Engineering specifications
4.6.1	Approved materials for ongoing production
4.6.2	Subcontractor development Scheduling subcontractors
4.6.3	Restricted substances
4.9	Government safety and environmental regulations Designated special characteristics Preventive maintenance
4.9.1	Process monitoring and operator instructions
4.9.2	Preliminary process capability requirements
4.9.3	Ongoing process performance requirements
4.9.4	Modified preliminary or ongoing capability requirements
4.9.5	Verification of job set-ups
4.9.6	Process changes
4.9.7	Appearance items
4.10.1	Acceptance criteria Accredited laboratories
4.10.2	Incoming product quality
4.10.3	Defect prevention
4.10.4	Layout inspection and functional testing
4.11.3	Inspection, measuring and test equipment records
4.11.4	Measurement system analysis
4.12	Product layout Supplemental verification
4.13.1	Suspect product
4.13.3	Control of reworked product
4.13.4	Engineering approved product authorisation
4.14.1	Problem solving methods
4.14.2	Returned product test/analysis
4.15.3	Inventory
4.15.4	Customer packaging standards Labelling
4.15.6	Supplier delivery performance monitoring Production scheduling Shipment notification system

4.16	Record retention
	Superseded parts
4.17	Inclusion of working environment
4.18	Training as a strategic issue
4.19	Feedback of information form service
4.20.2	Selection of statistical tools
	Knowledge of basic statistical concepts
II-1.1	Production part approval
II-1.2	Engineering change validation
II-2.1	Continuous improvement
II-2.2	Quality and productivity improvements
II-2.3	Techniques for continuous improvement
II-3.1	Facilities, equipment and process planning and effectiveness
II-3.2	Mistake proofing
II-3.3	Tool design and fabrication
II-3.4	Tooling management

Registered Auditors

The figures in Tables B-6 and B-7 for the number of registered auditors world-wide are those of the IRCA[1] only and do not include auditors registered with the RBA or other organizations.

	UK	Rest of World	Total
Lead Auditors	2398	2590	4988
Auditors	556	992	1548
Provisional Auditors	420	564	984
Lead TickIT Auditors	45	13	58
Senior TickIT Auditors	18	12	30
Provisional TickIT Auditors	57	27	84
Lead Auditors (Aerospace)	80	26	106
Auditors (Aerospace)	6	1	7
Provisional Auditors (Aerospace)	9	2	11
Lead Auditors (Pharmaceutical supplier)	18	2	20
Auditors (Pharmaceutical supplier)	1	2	3
Provisional Auditors (Pharmaceutical supplier)	1	0	1
Provisional Auditors (Maritime)	2	0	2
Internal Auditors	294	159	453
Total Registered	**3905**	**4390**	**8295**

Table B-6 Registered auditors by category

	UK	Rest of World	Total
Lead Auditors	2541	2631	5172
Auditors (except Internal Auditors)	581	1007	1588
Provisional Auditors	489	593	1082
Total Registered 2nd & 3rd Party	**3611**	**4231**	**7842**

Table B-6 Registered auditors - all categories

[1] Source: IRCA March 1996

Country	No	Country	No	Country	No
Argentina	7	India	242	Russia	16
Australia	279	Indonesia	26	Saudi Arabia	20
Austria	9	Ireland	103	Serbia	1
Bahrain	3	Israel	9	Singapore	244
Belgium	30	Italy	46	Seychelles	1
Brazil	101	Iran	1	Slovakia	1
Brunei	1	Japan	636	Slovenia	5
Bulgaria	1	Jordan	1	South Africa	30
Canada	73	Kenya	1	Spain	48
Chile	6	Korea	224	Sri Lanka	13
China	42	Kuwait	1	Sweden	40
Columbia	0	Malaysia	72	Switzerland	16
Croatia	7	Mauritius	14	Taiwan	57
Cuba	1	Mexico	12	Tanzania	1
Cyprus	12	Morocco	1	Thailand	26
Czech Republic	6	Netherlands	129	Trinidad & Tobago	2
Denmark	31	New Zealand	68	Tunisia	5
Egypt	14	Oman	9	Turkey	13
Finland	10	Norway	42	United Arab Emirates	20
Fiji	1	Pakistan	1	United Kingdom	3611
France	45	Philippines	10	USA	1100
Germany	44	Poland	10	Venezuela	4
Greece	15	Portugal	4	Yugoslavia	1
Hong Kong	136	Puerto Rico	1	Zimbabwe	4
Hungary	12	Romania	13		

Table B-8 Registered auditors by country (excluding internal auditors)

Auditor registration bodies

At an international conference in Cairns, Australia in July 1995, agreement was reached to create the International Auditor and Training Certification Association (IATCA). The purpose of IATCA is to harmonize auditor registration criteria so that auditors need only register with one auditor registration body to be accepted by others. This particularly affects auditors who frequently work in more than one country and who are currently subject to the requirements for multiple registration.

The IATCA is not an auditor registration body itself. Currently the IATCA is an association of national auditor registration bodies that have expressed an intent to comply in organizational terms to the IATCA criteria. The IATCA comes into operation early in 1996. Auditors will register with one of the accredited auditor registration bodies and those who satisfy the IATCA criteria and who wish to register additionally as IATCA auditors may do so.

Thirteen auditor registration bodies were signatories to a Memorandum of Understanding (MoU), indicating their intention to adopt the IATCA criteria for auditor certification and auditor training. The signatories of the MoU are:

1   Joint Accreditation System of Australia and New Zealand (JAS-ANZ)

2   Quality Society of Australasia (QSA)

3   National Institute of Metrology, Standardization and Industrial Quality (ANI-MATOR), Brazil

4   China Council for the Quality System (ISO 9000) of the Export Manufacturers (CCQSEM)

5   China National Registration Board for Auditors (CRBA/CSBTS)

6   Institute for Certification of Auditors (ICA) France

7   The Japan Accreditation Board for Quality System Registration (JAB)

8   Quality and Environmental Management Systems Accreditation Division, Korea Industrial Advancement Administration (KIAA)

9   Singapore Institute of Standards and Industrial Research (SISIR)

10 Southern African Council for the Certification of Quality System Auditors (SACCQA)

11 Bureau of Commodity Inspection and Quarantine, Ministry of Economic Affairs, Chinese Taipei

12 International Register of Certificated Auditors (IRCA)

13 Registrar Accreditation Board (RAB)

Appendix C

Requirements of ISO 9001

4.1 Management responsibility

4.1.1 Quality policy

1   Define and document policy for quality.

2   Define and document objectives for quality.

3   Define and document commitment to quality.

4   Cause the policy to be relevant to organizational goals.

5   Cause the policy to be relevant to needs of customers.

6   Cause the policy to be understood at all levels.

7   Implement the quality policy at all levels.

8   Maintain the quality policy at all levels.

4.1.2 Organization

4.1.2.1 Responsibility and authority

9   Define and document the responsibility and authority of managers.

10   Define and document the responsibility and authority of workers.

11   Define and document the responsibility and authority of verification staff.

12   Define and document the responsibility and authority of staff who prevent the occurrence of nonconformities.

13   Define and document the responsibility and authority of staff who identify and record problems relating to products, processes and the quality system.

14   Define and document the responsibility and authority of staff who provide solutions to problems.

15   Define and document the responsibility and authority of staff who verify the implementation of solutions.

16   Define and document the responsibility and authority of personnel who control the processing, delivery or installation of nonconforming product.

4.1.2.2 Resources

17   Identify resource requirements.

18   Provide adequate resources.

19   Assign trained personnel to management activities.

20   Assign trained personnel to performance of work.

21   Assign trained personnel to verification activities.

4.1.2.3 Management Representative

22   Appoint a member of the supplier's own management.

23   Delegate authority for ensuring that the quality system is established.

24   Delegate authority for ensuring that the quality system is implemented.

25   Delegate authority for ensuring that the quality system is maintained.

26   Delegate authority for reporting on the performance of the quality system.

4.1.3 Management review

27   Review the quality system at defined intervals.

28   Establish that the quality system continues to be suitable and effective in satisfying ISO 9001.

29   Establish that the quality system continues to be suitable and effective in satisfying the stated policy and objectives.

30   Record the results of management reviews.

4.2 Quality system

4.2.1 General

31   Establish a quality system.

32   Document a quality system.

33   Maintain a documented quality system.

34   Prepare a Quality Manual covering the requirements of the standard.

35   Include or make reference in the Quality Manual to the procedures used in the quality system.

36   Outline the structure of the documentation used in the quality system in the Quality Manual.

4.2.2 Quality system procedures

37   Prepare documented procedures consistent with the requirements of the standard.

38   Prepare documented procedures consistent with the requirements of the quality policy.

39   Effectively implement the quality system and its documented procedures.

40   Ensure that the range and detail of procedures is dependent upon the complexity of work, the methods used, and the skills and training needed.

4.2.3 Quality planning

41   Define and document how requirements for quality will be met.

42   Ensure that quality planning is consistent with all other requirements of the quality system.

43   Document quality planning in a format to suit the method of operation.

44   Give consideration to the preparation of quality plans.

45   Give consideration to the identification and acquisition of any control needed.

46   Give consideration to the identification and acquisition of any processes needed.

47   Give consideration to the identification and acquisition of any equipment needed.

48   Give consideration to the identification and acquisition of any fixtures needed.

49   Give consideration to the identification and acquisition of any resources needed.

50   Give consideration to the identification and acquisition of any skills needed.

51   Give consideration to ensuring the compatibility of the design and applicable documentation with specified requirements.

52   Give consideration to the compatibility of the production, installation and servicing process and applicable documentation with specified requirements.

53   Give consideration to the compatibility of inspection and test procedures with specified requirements.

54   Give consideration to the updating of quality control techniques.

55   Give consideration to the updating of inspection and test techniques.

56   Give consideration to the development of new instrumentation.

57   Give consideration to the identification of any measurement capabilities.

58   Give consideration to the identification of appropriate verification.

59   Give consideration to the clarification of standards of acceptability.

60   Give consideration to the identification of quality records.

61   Give consideration to the preparation of quality records.

4.3 Contract review

4.3.1 General

62   Establish and maintain documented procedures for contract review.

63   Establish and maintain procedures for the co-ordination of contract review activities.

4.3.2 Review

64   Review tenders before submission.

65   Review contracts before acceptance.

66   Ensure that requirements are adequately defined and documented before acceptance.

67   Ensure that order requirements are agreed before acceptance.

68   Ensure that differences between tender and contract/order requirements are resolved before acceptance.

69   Ensure capability to meet contract/order requirements before tender submission or contract acceptance.

4.3.3 Amendment to a contract

70   Identify how contract amendments are made.

71   Identify how contract amendments are correctly transferred to the functions concerned.

4.3.4 Records

72   Maintain records of contract reviews.

4.4 Design control

4.4.1 General

73   Establish and maintain documented procedures to control the design of product.

74   Establish and maintain documented procedures to verify the design of product.

4.4.2 Design and development planning

75   Prepare plans for each design and development activity.

76   Describe or reference design and development activities in the design and development plans.

77   Define the responsibility for design and development activities in the design and development plans.

78   Assign design and development activities to qualified personnel.

79   Equip design and development personnel with adequate resources.

80   Update design and development plans as the design evolves.

4.4.3 Organizational and technical interfaces

81   Define and document the organizational interfaces between groups which input into the design process.

82   Define and document the technical interfaces between groups which input into the design process.

83   Transmit organizational interface information to groups which input to the design process.

84   Transmit technical interface information to groups which input to the design process.

85   Regularly review organizational interface information.

86   Regularly review technical interface information.

4.4.4 Design input

87   Identify and document design input requirements relating to the product.

88   Identify and document statutory and regulatory requirements relating to the product.

89   Review for adequacy the selection of design input information.

90   Resolve incomplete, ambiguous or conflicting requirements with those responsible.

91   Take into consideration the results of contract reviews in documenting design input requirements.

4.4.5 Design output

92   Document design output.

93   Express design output in terms that can be verified and validated against design input requirements.

94   Ensure that design output meets design input requirements.

95   Ensure design output contains or makes reference to acceptance criteria.

96   Identify characteristics that are crucial to the safe and proper functioning of the product.

97   Review design output documents before release.

4.4.6 Design review

98   Plan formal documented reviews of the design results at appropriate stages.

99   Conduct formal documented reviews of the design results at appropriate stages.

100   Include representatives of all functions concerned with the design stage at each design review.

101   Include specialist personnel as required at each design review.

102   Maintain records of each design review.

4.4.7 Design verification

103 Perform design verification at appropriate stages.

104 Ensure that design stage output meets design stage input.

105 Record design verification measures.

4.4.8 Design validation

106 Perform design validation.

107 Ensure that product conforms to defined user needs and/or requirements.

4.4.9 Design changes

108 Identify all design changes.

109 Identify all modifications.

110 Document all design changes.

111 Document all modifications.

112 Review all design changes before their implementation.

113 Ensure all design changes are approved by authorized personnel before their implementation.

4.5 Document control

4.5.1 General

114 Establish and maintain document control procedures for documents of internal origin.

115 Control all documents that relate to the requirements of the standard.

116 Establish and maintain data control procedures.

117 Control all data that relate to the requirements of the standard.

118 Establish and maintain document control procedures for documents of external origin.

119 Control all documents of external origin that relate to the requirements of the standard.

4.5.2 Document and data approval and issue

120   Review documents for adequacy prior to release.

121   Review data for adequacy prior to release.

122   Ensure that documents are approved by authorized personnel prior to release.

123   Ensure that data are approved by authorized personnel prior to release.

124   Establish and maintain a master list or document control procedure for identifying the current revision status of documents.

125   Ensure that the master list or document control procedure is readily available.

126   Preclude the use of invalid and/or obsolete documents.

127   Ensure that the pertinent issues of appropriate documents are available at all essential locations.

128   Prevent unintended use of invalid and/or obsolete documents.

129   Identify any retained obsolete documents.

4.5.3 Document and data changes

130   Ensure that changes to documents are reviewed by the same functions that performed the original review.

131   Ensure that changes to documents are approved by the same functions that performed the original approval.

132   Provide the designated functions with access to pertinent background information upon which to base their review and approval.

133   Identify the nature of change in the document or appropriate attachment.

4.6 Purchasing

4.6.1 General

134   Establish and maintain documented purchasing procedures.

135   Ensure that purchased product conforms to specified requirements.

4.6.2 Evaluation of subcontractors

136   Evaluate subcontractors.

137 Select subcontractors on the basis of their ability to meet subcontract requirements.

138 Define the type and extent of control exercised over subcontractors.

139 Base subcontractor controls upon the type of product.

140 Base subcontractor controls upon the impact of subcontract product on the
quality of the final product.

141 Base subcontractor controls on the quality audit reports and/or quality records
of the previously demonstrated capability and performance of subcontractors.

142 Establish and maintain quality records of acceptable subcontractors.

4.6.3 Purchasing data

143 Ensure that purchasing documents clearly describe the product ordered.

144 Describe the type, class, grade of product ordered in purchasing documents.

145 In purchasing documents define the identity and applicable issues of relevant
technical data for product ordered.

146 In purchasing documents define requirements for approval or qualification of
product, procedures, process equipment and personnel.

147 In purchasing documents define the title, number and issue of the quality system
standard to be applied.

148 Review and approve purchasing documents for adequacy of specified requirements prior to release.

4.6.4 Verification of purchased product

4.6.4.1 *Supplier verification at subcontractor's premises*

149 Specify verification arrangements in purchasing documents where purchased
product is to be verified at subcontractor's premises.

150 Specify method of product release in purchasing documents where purchased
product is to be verified at subcontractor's premises.

4.6.4.2 *Customer verification of subcontracted product*

151 Afford customers the right to verify at subcontractor's premises that subcontract
product conforms to specified requirements.

4.7 Control of customer supplied product

152 Establish and maintain documented procedures for controlling the verification of customer supplied product.

153 Establish and maintain documented procedures for controlling the storage of customer supplied product.

154 Establish and maintain documented procedures for controlling the maintenance of customer supplied product.

155 Record any customer supplied product that is lost, damaged or otherwise unsuitable for use.

156 Report to the customer any customer supplied product that is lost, damaged or otherwise unsuitable for use.

4.8 Product identification and traceability

157 Establish and maintain documented procedures where appropriate for identifying the product from receipt and during all stages of production, delivery and installation.

158 Establish and maintain documented procedures when specified for unique identification of individual product or batches.

159 Record the unique identification of product or batches.

4.9 Process control

160 Identify and plan the production, installation and servicing processes which directly affect quality.

161 Carry out processes directly affecting quality under controlled conditions.

162 Document those procedures which define the manner of production, installation and servicing.

163 Use suitable production, installation and servicing equipment.

164 Use a suitable working environment.

165 Comply with reference standards/codes, quality plans and/or documented procedures.

166 Monitor and control suitable process parameters.

167 Monitor and control suitable product characteristics.

168 Approve production, installation and servicing processes.

169 Approve production, installation and servicing equipment.

170 Stipulate criteria for workmanship in the clearest practical manner.

171 Provide suitable maintenance of equipment to ensure continuing process capability.

172 Ensure continuing process capability.

173 Use only qualified operators for special processes or employ continuous monitoring of process parameters.

174 Specify requirements for any qualification of process operations, equipment and personnel.

175 Maintain records for qualified processes, equipment and personnel.

4.10 Inspection and testing

4.10.1 General

176 Establish and maintain documented procedures for inspecting and testing activities to verify that specified requirements for the product are met.

177 Detail the required inspection and testing in the quality plan or documented procedures.

178 Detail the inspection and test records to be established in the quality plan or documented procedures.

4.10.2 Receiving inspection and testing

179 Ensure that incoming product is not used or processed until it has been verified as conforming to specified requirements.

180 Conduct receiving verification in accordance with the quality plan or documented procedures.

181 Give consideration to the amount of control exercised at subcontractor premises and the recorded evidence of conformance provided when determining the amount and nature of receiving inspection.

182 Positively identify product which is to be released for urgent production purposes prior to receipt verification.

183 Record products released for urgent production purposes prior to receipt verification.

4.10.3 In-process inspection and testing

184 Inspect and test product as required by quality plan and/or documented procedures.

185 Hold product until the required inspection and tests have been completed or reports have been received and verified.

4.10.4 Final inspection and testing

186 Carry out final inspection and testing in accordance with the quality plan and/or documented procedures.

187 Require all specified inspection and tests to be carried out and results to meet specified requirements.

188 Hold dispatch of product until all activities specified in the quality plan and/or documented procedures have been satisfactorily completed.

189 Hold dispatch of product until all the associated data and documentation is available and authorized.

4.10.5 Inspection and test records

190 Establish and maintain records which provide evidence that product has been inspected and/or tested.

191 Ensure that records show clearly whether the product has passed or failed the inspection criteria.

192 Subject failed product to procedures for control of nonconforming product.

193 Record the identify of the inspection authority responsible for release of product.

4.11 Inspection, measuring and test equipment

4.11.1 General

194 Establish and maintain documented procedures to control inspection, measuring and test equipment.

195 Establish and maintain documented procedures to calibrate inspection, measuring and test equipment.

196 Establish and maintain documented procedures to maintain inspection, measuring and test equipment.

197 Establish and maintain documented procedures to control test software.

198 Establish and maintain documented procedures to calibrate test software.

199 Establish and maintain documented procedures to maintain test software.

200 Use inspection, measuring and test equipment in a manner which ensures that the measurement uncertainty is known.

201 Ensure that measurement uncertainty is consistent with the required measurement capability.

202 Check comparative references to prove that they are capable of verifying the acceptability of product prior to release for use during production, installation and servicing.

203 Recheck comparative references at prescribed intervals.

204 Establish the extent and frequency of checks on comparative references.

205 Maintain records of checks on comparative references.

206 Cause technical data of inspection, measuring and test equipment to be available to the customer when required.

4.11.2 Control procedure

207 Determine the measurements to be made and the accuracy required to demonstrate conformance of product to specified requirements.

208 Select the appropriate inspection, measuring and test equipment that is capable of the necessary accuracy and precision to demonstrate conformance of product to specified requirements.

209 Identify all inspection, measuring and test equipment that can affect product quality.

210 Calibrate and adjust all inspection, measuring and test equipment that can affect product quality.

211 Calibrate devices at prescribed intervals or prior to use.

212 Calibrate devices against certified equipment having a known relationship to internationally or nationally recognized standards.

213 Document the basis of calibration where no recognized standards exist.

214 Define the process employed for the calibration of inspection, measuring and test equipment.

215 Define equipment type and unique identification.

216 Define the location of equipment.

217 Define the frequency of checks and the check method.

218 Define the acceptance criteria.

219 Define the action to be taken when results are unsatisfactory.

220 Identify inspection, measuring and test equipment with a suitable indicator or approved identification record to show the calibration status.

221 Maintain calibration records for inspection, measuring and test equipment.

222 Assess and document the validity of previous inspection and test results when inspection, measuring and test equipment is found to be out of calibration.

223 Carry out calibration in a suitable environment.

224 Carry out inspection, measurements and tests in a suitable environment.

225 Maintain the accuracy and fitness for use of inspection, measuring and test equipment.

226 Safeguard inspection, measuring and test facilities from adjustments which would invalidate the calibration setting.

4.12 Inspection and test status

227 Identify the inspection and test status of product by means which indicate conformance or nonconformance with inspections and test performed.

228 Maintain the inspection and test status as defined in the quality plan and/or documented procedures throughout production.

229 Maintain the inspection and test status as defined in the quality plan and/or documented procedures throughout installation.

230 Maintain the inspection and test status as defined in the quality plan and/or documented procedures throughout servicing.

231 Only dispatch, use or install product which has passed the required inspections and tests unless released under authorized concession.

4.13 Control of nonconforming product

4.13.1 General

232 Establish and maintain documented procedures to prevent unintended use of nonconforming product.

233 Provide for the identification of nonconforming product.

234 Provide for the documentation of nonconforming product.

235 Provide for the evaluation of nonconforming product.

236 Provide for the segregation of nonconforming product.

237 Provide for the disposition of nonconforming product.

238 Notify functions concerned of nonconforming product.

4.13.2 Nonconformity review and disposition

239 Define the responsibility for the review and authority for the disposition of nonconforming product.

240 Review nonconforming product in accordance with documented procedures.

241 Report proposed use of nonconforming product to customer when required.

242 Record the actual condition of accepted nonconformity and repairs.

243 Re-inspect repaired and/or reworked product in accordance with quality plan and/or documented procedures.

4.14 Corrective and preventive action

4.14.1 General

244 Establish and maintain documented procedures for implementing corrective actions.

245 Establish and maintain documented procedures for implementing preventive actions.

246 Eliminate the causes of actual nonconformities to a degree appropriate to the magnitude of problems and commensurate with the risks encountered.

247 Eliminate the causes of potential nonconformities to a degree appropriate to the magnitude of problems and commensurate with the risks encountered.

248 Implement any changes to the documented procedures resulting from corrective actions.

249 Implement any changes to the documented procedures resulting from preventive actions.

250 Record any changes to the documented procedures resulting from corrective actions.

251 Record any changes to the documented procedures resulting from preventive actions.

4.14.2 Corrective action

252 Provide for the effective handling of customer complaints.

253 Provide for the effective handling of reports of product nonconformities.

254 Investigate the cause of nonconformities relating to products.

255 Investigate the cause of nonconformities relating to processes.

256 Investigate the cause of nonconformities relating to the quality system.

257 Record the result of investigations to determine the cause of nonconformities.

258 Determine the corrective actions needed to eliminate the cause of nonconformities.

259 Apply controls to ensure that corrective action is taken and is effective.

4.14.3 Preventive action

260 Use appropriate sources of information to detect, analyse and eliminate potential causes of nonconformities.

261 Determine the steps needed to deal with any problems requiring preventive action.

262 Initiate preventive action.

263 Apply controls to ensure that preventive action is effective.

264 Submit relevant information on preventive actions taken for management review.

4.15 Handling, storage, packaging, preservation and delivery

4.15.1 General

265 Establish and maintain documented procedures for handling of product.

266 Establish and maintain documented procedures for storage of product.

267 Establish and maintain documented procedures for packaging of product.

268 Establish and maintain documented procedures for preservation of product.

269 Establish and maintain documented procedures for delivery of product.

4.15.2 Handling

270 Provide methods of handling product that prevent damage and deterioration.

4.15.3 Storage

271 Use designated storage areas or stock rooms to prevent damage and deterioration of product pending use or delivery.

272 Stipulate appropriate methods for authorizing receipt into storage areas.

273 Stipulate appropriate methods for authorizing dispatch from storage areas.

274 Assess the condition of product in stock at appropriate intervals.

4.15.4 Packaging

275 Control packing processes to the extent necessary to ensure conformance to specified requirements.

276 Control packaging processes to the extent necessary to ensure conformance to specified requirements.

277 Control marking processes to the extent necessary to ensure conformance to specified requirements.

278 Control the material used in packing processes to the extent necessary to ensure conformance to specified requirements.

4.15.5 Preservation

279 Apply appropriate methods of preservation of product when the product is under the supplier's control.

280 Apply appropriate methods of segregation of product when the product is under the supplier's control.

4.15.6 Delivery

281 Arrange for the protection of product after final inspection and test and up to delivery when required.

4.16 Control of quality records

282 Establish and maintain documented procedures for the identification of quality records.

283 Establish and maintain documented procedures for the collection of quality records.

284 Establish and maintain documented procedures for the indexing of quality records.

285 Establish and maintain documented procedures for the access of quality records.

286 Establish and maintain documented procedures for the filing of quality records.

287 Establish and maintain documented procedures for the storage of quality records.

288 Establish and maintain documented procedures for the maintenance of quality records.

289 Establish and maintain documented procedures for the disposition of quality records.

290 Maintain quality records to demonstrate conformance to specified requirements.

291 Maintain quality records to demonstrate the effective operation of the quality system.

292 Maintain subcontractor quality records.

293 Ensure that quality records are legible.

294 Store quality records in such a way as to prevent damage, deterioration and loss.

295 Retain quality records in such a way that they are readily retrievable.

296 Establish and record retention times for quality records.

297 Cause quality records to be available for evaluation by the customer for an agreed period when required.

4.17 Internal quality audits

298 Establish and maintain documented procedures for planning internal quality audits.

299 Establish and maintain documented procedures for implementing internal quality audits.

300 Verify that quality activities and related results comply with planned arrangements by conducting internal quality audits.

301 Determine the effectiveness of the quality system by conducting internal quality audits.

302 Schedule internal quality audits on the basis of the status and importance of the activity to be audited.

303 Carry out internal quality audits using personnel who are independent of those having a direct responsibility for the activity being audited.

304 Record the results of internal quality audits.

305 Bring the results of internal quality audits to the attention of the personnel having responsibility in the area audited.

306 Take timely corrective action on deficiencies found during audits.

307 Verify the implementation of corrective actions by follow-up audits.

308 Verify the effectiveness of corrective actions by follow-up audits.

4.18 Training

309 Establish and maintain documented procedures for identifying training needs.

310 Provide for the training of all personnel performing specific assigned activities affecting quality.

311 Qualify personnel performing specific assigned tasks on the basis of appropriate education, training and/or experience as required.

312 Maintain appropriate records of training.

4.19 Servicing

313 Establish and maintain documented procedures for performing servicing.

314 Establish and maintain documented procedures for verifying that servicing meets specified requirements.

315 Establish and maintain documented procedures for reporting that servicing meets specified requirements.

4.20 Statistical techniques

4.20.1 Identification of need

316 Identify the need for statistical techniques required for establishing process capability.

317 Identify the need for statistical techniques required for controlling process capability.

318 Identify the need for statistical techniques required for verifying process capability.

319 Identify the need for statistical techniques required for establishing product characteristics.

320 Identify the need for statistical techniques required for controlling product characteristics.

321 Identify the need for statistical techniques required for verifying product characteristics.

4.20.2 Procedures

322 Establish and maintain documented procedures to implement the application of the identified statistical techniques.

323 Establish and maintain documented procedures to control the application of the identified statistical techniques.

Appendix D

Auditor's dictionary

The terms in this dictionary have been taken from my *ISO 9000 Quality Systems Handbook* and extended to meet the needs of auditors. They do not repeat definitions given in ISO 8402 but instead provide less verbose explanations so as to help auditors, especially those born outside the UK.

Acceptance criteria The standard against which a comparison is made to judge conformance.

Accreditation A process by which organizations are authorized to conduct certification of conformity to prescribed standards. See *Certification body* or *Registrar*.

Activities affecting quality Any activity which affects the determination of product or service characteristics, their specification, achievement or verification or means to plan organize, control, assure or improve them.

Adequacy audit An audit carried out to establish that the quality system documentation adequately addresses the requirements of a prescribed standard. N.B. also referred to as a *Documentation audit*.

Adequate Adequate means suitable for the purpose. The term 'adequate' appears several times in the standard allowing the auditor to vary the criteria for adequacy and hence not use a finite process to verify that the requirements have been met.

Aggressive behaviour Behaviour likely to intimidate the auditee and reduce their co-operation. A person who forces his/her opinions or demands upon others and offers no co-operation.

Appropriate

Appropriate means appropriate to the circumstances and requires knowledge of these circumstances. Without criteria, an auditor is left to decide what is or is not appropriate.

Approved

Approved means that it has been confirmed as meeting the requirements.

Assertive behaviour

Behaviour likely to cause positive results without intimidating the auditee. A person who offers resistance when attempts are made to distract him/her from their objective.

Assessment

The act of determining the extent of compliance with requirements.

Assessment period

The time, usually in months, between full assessments.

Assurance

Evidence (verbal or written) that gives confidence that something will or will not happen or has or has not happened.

Audit

An examination of records or activities to verify their accuracy usually by someone other that the person responsible for them. A modifying adjective is usually used for specific types of audit e.g. certification audit, quality system audit.

Audit brief

A statement defining the boundary conditions and requirements for an audit. Specifically, details of the organization to be audited, the audit objective and its scope.

Audit objectives

The result which is to be achieved by the audit and usually one or more of the following:

- Whether certain agreed provisions, if implemented, will yield the required results

- Whether only the agreed provisions are being implemented

- Whether the provisions have yielded results that are fit for their purpose and meet the needs of those who require them

- Whether a certificate of conformance can be issued

- Whether some improvement is necessary before awarding a certificate

Audit plan

A chart showing the areas to be audited during a specific audit of an organization, including the timing and names of the auditors involved.

Audit programme

A chart showing the audits scheduled to be performed on specific organizations over a given period, usually 12 months.

Audit purpose

To establish, by an unbiased means, factual information on some aspect of performance.

Audit report

A factual account of the results of the audit including the good points, extent of compliance, nonconformities, conclusions, recommendations and corrective actions.

Audit scope

The range or extent of the audit including the standard or contract against which the audit is to be conducted, the products and services and the processes to be included e.g. design, development, production, installation or servicing.

Audit trail

A path of enquiry and discovery that an auditor follows in search of objective evidence.

Auditee

The person whose operations are being audited.

Auditor

A person who has the qualifications and is authorized to conduct an audit.

Auditors review meeting

A meeting convened by the Lead Auditor to review progress, discuss nonconformities, resolve problems and compile the audit report.

Authority

The right to take actions and make decisions.

Authorized

A permit to do something or use something which may not necessarily be approved.

Business process	A series of operations that are an essential part of a business.
Business process model	A diagram of the key processes that convert external inputs into outputs showing their interrelationships and channels along which product or information flows.
Business processes	The composite of all processes that define how an organization conducts its business.
Calibrate	To standardize the quantities of a measuring instrument.
Capability audit	An audit performed to verify that a process has the capability to consistently yield product which meets agreed requirements
Certification	A process by which a product, process, person or organization is deemed to meet specified requirements.
Certification audit	An audit performed for the purpose of certifying a product, process, person or organization.
Certification body	An organization that is authorized to certify organizations. The body may be accredited or non-accredited.
Check list (audit)	A list of topics or questions that guide an auditor conducting an audit. An aid to memory rather than a list of all questions to be asked.
Clause of the standard	A numbered paragraph or subsection of the standard containing one or more related requirements such as 4.10.3. N.B. Each item in a list is also a clause (see also *Quality system element*).
Client feedback meeting	A meeting convened by the Lead Auditor to report progress, resolve problems and obtain agreement to any nonconformities declared.
Closing meeting	A meeting between the auditor(s) and representatives of the organization audited convened to report and agree the results of the audit and to agree follow-up action.
Code of conduct	A set of rules that govern the behaviour of an auditor when conducting an audit.

Codes	A systematically arranged and comprehensive collection of rules, regulations or principles.
Commitment	An obligation a person or organization undertakes to fulfil i.e. doing what you say you will do.
Comparative reference	A standard used to determine differences between it and another entity.
Compliance audit	An audit performed to determine compliance with specified requirements. The term is sometimes limited to that part of an audit that verifies whether documented practices are being followed.
Concession	Permission granted by an acceptance authority to supply product or service that does not meet the prescribed requirements.
Conformance audit	An audit performed to determine conformance or conformity with specified requirements (see also *Compliance audit*).
Conforms to specified requirements	Meets the requirements that have been specified by the customer or the market.
Continual assessment	An assessment in which selected parts of the quality system are assessed on each visit and which over a given period subject the whole quality system to re-assessment.
Contract	An agreement formally executed by both customer and supplier (enforceable by law) which requires performance of services or delivery of products at a cost to the customer in accordance with stated terms and conditions.
Contractual requirements	Requirements specified in a contract.
Control	The act of preventing or regulating change in parameters, situations or conditions.
Control methods	Particular ways of providing control that do not constrain the sequence of steps in which the methods are carried out.
Control procedure	A procedure that controls product or information as it passes through an organization.

Controlled conditions	Arrangements which provide control over all factors that influence the result.
Corrective action	Action planned or taken to stop something from recurring.
Criteria for workmanship	Acceptance standards based on qualitative measures of performance.
Customer complaints	Any adverse report (verbal or written) received by a supplier from a customer.
Customer feedback	Positive or negative information received by a supplier from a customer.
Customer supplied product	Hardware, software, documentation or information owned by the customer which is provided to a supplier for use in connection with a contract and which is returned to the customer either incorporated in the supplies or at the end of the contract.
Data	Information that is organized in a form suitable for manual or computer analysis.
Define and document	To state in written form, the precise meaning, nature or characteristics of something.
Demonstrate	To prove by reasoning, objective evidence, experiment or practical application.
Design	A process of originating a conceptual solution to a requirement and expressing it in a form from which a product may be produced or a service delivered.
Design and development	Design creates the conceptual solution and development transforms the solution into a fully working model.
Design review	A formal documented and systematic critical study of a design by people other than the designer.
Desk top audit	A documentation audit performed at a desk with the organization's quality system documentation.
Disposition	The act or manner of disposing of something.
Documentation audit	An audit carried out to determine whether an organization's documented quality system makes adequate

provision for meeting the requirements of a given standard (see also *Adequacy Audit* or *Desk top audit*).

Documented procedures Procedures that are formally laid down in a reproducible medium such as paper or magnetic disk.

Effectiveness of the system The extent to which the (quality) system fulfils its purpose.

Ensure To make certain that something will happen.

Establish and maintain To set up an entity on a permanent basis and retain or restore it in a state in which it can fulfil its purpose or required function.

Evaluation To ascertain the relative goodness, quality or usefulness of an entity with respect to a specific purpose.

Evidence of conformance Documents which testify that an entity conforms with certain prescribed requirements.

Executive responsibility Responsibility vested in those personnel who are responsible for the whole organization's performance. Often referred to as top management.

Extended assessment An assessment in which those parts of an organization excluded from the full assessment sample are subject to audit on each visit together with system maintenance monitoring and a complete re-assessment at the end of each period.

External audits Audits carried out by an organization independent of the organization audited. Independence has to be such that there is no financial association other than a contract.

Final inspection and testing The last inspection or test carried out by the supplier before ownership passes to the customer.

Finding Information revealed from an examination of documentation, items or activities.

First party audits Audits of a company or parts thereof by personnel employed by the company. These audits are also called *Internal audits*.

Follow-up audit An audit carried out following and as a direct consequence of a previous audit to determine whether agreed actions have been taken and are effective.

Functions	In the organizational sense, a function is a special or major activity (often unique in the organization) which is needed in order for the organization to fulfil its purpose and mission. Examples of functions are design, procurement, personnel, manufacture, marketing, maintenance etc. Departments may perform one or more functions but a department is a component of the organization, not a function.
Guide	A person who escorts the auditor during the audit.
Hearsay evidence	Oral statements concerning a situation that has not been observed directly.
Identification	The act of identifying an entity, i.e. giving it a set of characteristics by which it is recognizable as a member of a group.
Implement	To carry out a directive.
Implementation audit	An audit carried out to establish whether actual practices conform to the documented quality system. N.B. also referred to as a *Conformance audit* or *Compliance audit*.
Importance of activities in auditing	The relative importance of the contribution an activity makes to the fulfilment of an organization's objectives.
In-process	Between the beginning and the end of a process.
Indexing	A means of enabling information to be located.
Inspection	The examination of an entity to determine whether it conforms to prescribed requirements.
Inspection authority	The person or organization who has been given the right to perform inspections.
Inspection, measuring and test equipment	Devices used to perform inspections, measurements and tests.
Installation	The process by which an entity is fitted into a larger entity.
Internal audits	See *First party audits*.
Issues of documents	The revision state of a document.

Lead auditor	A person qualified and authorized to lead an audit team.
Major nonconformity	The absence or total breakdown of the provisions required to cause product conformity or prevent product nonconformity with the expectations and needs of customers. Absence means a lack of adequate provisions in theory and in practice. A total breakdown means that adequate provisions are in place but they are currently not being implemented.
Manage work	To manage work means to plan, organize and control the resources (personnel, financial and material) and tasks required to achieve the objective for which the work is needed.
Management representative	The person management appoints to act on their behalf to manage the quality system. Their actual title is irrelevant.
Master list	An original list from which copies can be made.
Measurement capability	The ability of a measuring system (device, person and environment) to measure true values to the accuracy and precision required.
Measurement uncertainty	The variation observed when repeated measurements of the same parameter on the same specimen are taken with the same device.
Minor nonconformity	Any failure to meet one or more requirements of the standard that cannot be classified as a major nonconformity.
Modifications	Entities altered or reworked to incorporate design changes.
Monitoring	To check periodically and systematically. It does not imply that any action will be taken.
Nationally recognized standards	Standards of measure that have been authenticated by a national body.
Nature of change	The intrinsic characteristics of the change (what has changed and why).
Nonconformity	A failure to meet a specified requirement.

Objective	The result that is to be achieved usually by a given time.
Objective evidence	Findings that can be substantiated by information which is factual and which can be verified.
Obsolete documents	Documents that are no longer required for operational use. They may be useful as historic documents.
On-site audit	An audit performed on the auditee's premises.
Opening meeting	A meeting between the auditor(s) and representatives of the organization to be audited convened to confirm the arrangements prior to commencing the on-site audit.
Operating procedure	A procedure that describes how specific tasks are to be performed.
Organization audit	An audit performed to verify that the organization is structured and resourced to implement its stated policies and will achieve the stated objectives efficiently and effectively.
Organizational interfaces	The boundary at which organizations meet and affect each other, expressed by the passage of information, people, equipment, materials and the agreement to operational conditions.
Organizational goals	Where the organization desires to be, in markets, in innovation, in social and environmental matters, in competition and in financial health.
Passive behaviour	Behaviour of someone who offers no resistance and is likely to be lead by a more assertive person.
Passive-aggressive behaviour	Someone who offers neither resistance nor co-operation.
Periodic assessment	An assessment in which the quality system is subject to system maintenance monitoring between periods and complete re-assessment at the end of each period.
Plan	Provisions made to achieve an objective.
Planned arrangements	All the arrangements made by the supplier to achieve the customers requirements. They include the documented policies and procedures and the documents derived from such policies and procedures.

Planning audit	An audit performed to verify that the organization's plans or proposals for supplying a product or service will, if properly implemented, result in product or service that complies with specified requirements.
Policy	A guide to thinking, action and decision. Policies can exist at any level in an organization from corporate level to the lowest level where activities are performed.
Policy audit	An audit performed to verify that the documented policies of an organization promulgate the requirements of the market and the objectives of the business.
Positive recall	A means of recovering an entity by giving it a unique identity.
Positively identified	An identification given to an entity for a specific purpose that is both unique and readily visible.
Potential nonconformity	A situation which if left alone will in time result in a nonconformity.
Prevent	To stop something from occurring by a deliberate planned action.
Preventive action	Action proposed or taken to stop something from occurring.
Procedure	A sequence of steps to execute a routine activity. Procedures can address interdepartmental, departmental, process, group, section or individual activities.
Process	A sequence of tasks that combines the use of people, machines, methods, tools, environment, instrumentation and materials to convert given inputs into outputs of added value.
Process capability	The ability of a process to maintain product characteristics within pre-set limits.
Process parameters	Those variables, boundaries or constants of a process that restrict or determine the results.
Product	Anything produced by human effort, natural or man-made processes.

Production	The creation of products.
Purchaser	One who buys from another.
Purchasing documents	Documents that contain the supplier's purchasing requirements.
Qualification	Determination by a series of tests and examinations of products, related documents and processes that the product meets all the specified performance capability requirements.
Qualification approval	The status given to a supplier whose product has been shown to meet all the specified requirements.
Qualified personnel	Personnel who have been judged as having the necessary ability to carry out particular tasks.
Quality activities	Any activity that affects the ability of a product or service to satisfy stated or implied needs or the organization's ability to satisfy those needs. If the quality system defines the activities that need to be executed to achieve quality then any activity specified in the documented quality system is also a quality activity.
Quality conformance	The extent to which the product or service conforms with the specified requirements.
Quality costs	Costs incurred because failure is possible. The actual cost of producing an entity is the no-failure cost plus the quality cost. The no-failure cost is the cost of doing the right things right first time. The quality costs are the prevention, appraisal and failure costs.
Quality planning	Provisions made to prevent failure to satisfy customer needs and expectation and organizational goals.
Quality plans	Plans produced to define how specified quality requirements will be achieved, controlled, assured and managed for specific contracts or projects.
Quality problems	The difference between the achieved quality and the required quality.
Quality records	Objective evidence of the achieved features and characteristics of a product or service and the processes

applied to its development, design, production, installation, maintenance and disposal as well as records of assessments, audits and other examinations of an organization to determine its capability to achieve given quality requirements.

Quality requirements

Those requirements that pertain to the features and characteristics of a product or service which are required to be fulfilled in order to satisfy a given need.

Quality system

A tool for achieving, sustaining and improving quality. Such a system should integrate interconnected business processes which collectively cause the supply of conforming product/service and prevent the supply of nonconforming product/service.

Quality system assessments

External audits carried out by second or third parties.

Quality system element

A distinct part of the system which is governed by a set of requirements A subsection of the standard identified by a two-digit number such as 4.1, 4.2, 4.3 etc.

Quality system requirements

Requirements pertaining to the design, development, implementation and maintenance of quality systems.

Random failure

A failure which has a low probability of recurrence and which requires only remedial action to eliminate.

Registrar

See *Certification body*.

Registration

A process of recording details of organizations of assessed capability which have satisfied prescribed standards.

Regulatory requirements

Requirements established by law pertaining to products or services.

Related results

Results that arise out of performing an activity or making a decision. In the context of quality activities they may be documents, records, approval and acceptance decisions, disapproval and reject decisions, products, processes.

Remedial action

Action proposed or taken to remove a nonconformity (see also *Corrective action* and *Preventive action*). The action applies to the affected item, process or activity.

Representative sample	A sample of product or service which possesses all the characteristics of the batch from which it was taken.
Requirement of the standard	A sentence containing the word 'shall'. N.B. Some sentences contain multiple requirements such as 'To establish, document and maintain ...' This is in fact three requirements.
Responsibility	An area in which one is entitled to act on one's own accord.
Review	Another look at something.
Second party audits	Audits carried out by customers upon their suppliers.
Self audit	An audit carried out by a person responsible for the activities audited.
Service	Results that do not depend on the provision of products.
Service reports	Reports of servicing activities.
Servicing	Action to restore or maintain an item in an operational condition.
Shall	A provision that is binding.
Should	A provision that is optional.
Specified requirements	a) Requirements prescribed by the customer in a contract or b) Requirements prescribed by the supplier in a market requirement or design brief as a result of an analysis of the market need.
Status	The relative condition, maturity or quality of something.
Status of an activity	The maturity or relative level of performance of an activity to be audited.
Strategic quality audit	An audit performed to verify that the strategic plans of the organization address current and future legal, environmental, safety and market quality requirements.
Subcontract requirements	Requirements placed on a subcontractor which are derived from requirements of the main contract.

Subcontractor	A person or company that enters into a subcontract and assumes some of the obligations of the prime contractor.
Supplier	A person or company who supplies products or services to a purchaser.
Supplier approval audits	See *Vendor audits*.
Surveillance (quality system)	An activity performed to verify that the organization has maintained its quality system and it continues to be suitable for achieving its stated objectives and effective in providing an adequate degree of control over the organization's operations.
System audit	An audit carried out to establish whether the quality system conforms to a prescribed standard in both its design and its implementation.
System effectiveness	The ability of a system to achieve its stated purpose and objectives.
Systematic failure	A failure that has a high probability of recurrence due to inadequacy in the system and for which corrective action can be specified to eliminate the cause and prevent recurrence.
Technical interfaces	The physical and functional boundary between products or services.
Tender	A written offer to supply products or services at a stated cost.
Third party audits	External audits carried out by personnel who are neither employees of the customer nor the supplier and are usually employees of certification bodies or registrars.
Traceability	The ability to trace the history, application, use and location of an individual article or its characteristics through recorded identification numbers.
Unique identification	An identification that has no equal.
Validation	A process for establishing whether an entity will fulfil the purpose for which it has been selected or designed.
Vendor audits	An external audit of a supplier by its customers.

Verification	The act of establishing the truth or correctness of a fact, theory, statement or condition.
Verification activities	A special investigation, test, inspection, demonstration, analysis or comparison of data to verify that a product or service or process complies with prescribed requirements.
Verification requirements	Requirements for establishing conformance of a product or service with specified requirements by certain methods and techniques.
Work instructions	Instructions that prescribe work to be executed, who is to do it, when it is to start and be complete and how, if necessary, it is to be carried out.

Index

Contact details and disk offer

Comment, questions and constructive criticism may be sent via E-mail on the Internet to hoyle@celtic.co.uk or by letter to the address below.

In order to save you time in creating your own forms and check lists, the author has prepared a disk containing the check lists and forms in the book together with a more detailed ISO 9001 check list. Unlike the normal requirement check list used by the certification bodies, this check list contains questions that an auditor can actually ask to establish compliance with the requirements of each clause of the standard.

The price of the author's disk is £14.95 inc. VAT and P&P for mailing within the UK. The price for mailing outside the UK is £19.95, payable by cheque drawn on an overseas bank in the appropriate currency.

To obtain your disk, send a cheque with order, stating disk format, to:

  David Hoyle
  Royal Monnow
  Redbrook Road
  MONMOUTH
  Monmouthshire
  NP5 3LY
  United Kingdom

Please allow 21 days for delivery within the UK. Overseas orders will be dispatched by air mail.